JAMES MILLER

THE DAM BUILDERS

POWER FROM THE GLENS

First published in 2002 by
Birlinn Limited
West Newington House
10 Newington Road
Edinburgh
EH9 1QS

www.birlinn.co.uk

Reprinted 2003 (twice)

ISBN 1 84158 225 5

British Library Cataloguing-in-Publication Data
A catalogue record for this book is available from the British Library.

Series design: James Hutcheson
Layout and page make up: Mark Blackadder

Printed and bound in Great Britain by the Bath Press, Glasgow.

THE DAM BUILDERS

Contents

INTRODUCTION 7
ACKNOWLEDGEMENTS 10
PICTURE CREDITS 12

1. '... the most hopeful thing ...' 13

2. '... a cold job it was ...' 54

3. '... Bonus, Bonus ...' 123

4. 'You wouldn't call the King your uncle' 148

5. '... power and light have come to us ...' 182

6. '... still a huge part to play ...' 213

NOTES AND REFERENCES 250
MAIN SOURCES 253
APPENDIX 254

Introduction

Glen Cannich extends westward from the village of the same name in the heart of Inverness-shire. In the 1920s part of it belonged to the Chisholm estate, where the gamekeeper was Archie Chisholm. It was a remote and beautiful place, a long trough between ranges of hills, unfrequented by many visitors except for a brief period late every summer when the estate owners and their friends came for the hunting and fishing. It was the custom for tradesmen to come up from Inverness in the spring to do maintenance and prepare the buildings for occupancy. In one particular spring, Archie's wife, Margaret, said she would go along to open the lodge and check that the bothy where the tradesmen would stay was ready for their arrival. As she rounded a corner on the approach to the lodge she heard the sound of hammering, of boards being thrown down, of workmen talking, and, being a shy person, she turned back for home. When she told her husband the men had arrived, he said that was strange, they weren't expected until the next day, and went along to check. He found the place to be absolutely deserted. Margaret Chisholm was to maintain for the rest of her life that the noises she heard were a premonition of the work to come, for the Mullardoch dam stands now where she had this uncanny experience. There are many strands to the story of the building of the hydro-electric schemes in the Highlands after the Second World War. Mrs Chisholm's story typifies one strand and also shows how folklore thrives and still seeks to grace the more prosaic facts surrounding an historical event with an indication of their importance in the people's story.

The founding of the North of Scotland Hydro-Electric Board in 1943 and the subsequent construction of hydro-electric schemes and electrification of virtually all of the country north of the Highland Line was a major event in Highland history in the latter half of the twentieth century. A full account of the technical and political aspects of the great endeavour can be found in Peter Payne's *The Hydro*, published in 1988. It is in many respects an official history, done by Professor Payne with full access to Board archives. The story in this book is more of a people's history: it is very much how the momentous achievement of Highland electrification has been remembered by the men and women who worked on the schemes, and how it was recorded in local newspapers and other sources at the time. I have tried, therefore, to compliment Professor Payne's account in *The Hydro,* but have repeated enough of the official story to provide the political and historical framework in which the coming of 'the electric', as it was termed in everyday speech, took place.

My own memory of the time when electricity reached the village in Caithness where I was born and raised is fragmentary: I was only four or five years old, but I can recall the sudden appearance of street lamps. We lived in a house where we used Tilley paraffin lamps for illumination and a peat-fired range for cooking; both were then standard fixtures in Highland homes.

The building of the hydro-electric schemes in the Highlands in the three decades after the Second World War – the Hydro Board's major construction schemes came to an end in 1975 – is a story about rock and cement, frost, sweat and grease, and hard physical labour in a beautiful landscape. There is an heroic aspect to it, as nothing on such a scale had been attempted before; and the dams and tunnels came to symbolise far more than huge devices simply for the generation of electricity.

All the schemes came to be built in the belt of rugged mountain country that extends in a northward sweep from the Kintyre peninsula to the rolling vastness of Sutherland. The watershed dividing the burns and rivers that flow to the Atlantic from those that eventually disgorge into the North Sea lies close to the western side of the mainland. The Highlands are therefore scored by a series of long glens that lie with their heads within a few miles of the western seaboard but whose drainage is eastward. The catchment areas of the Tay, the Spey, the Findhorn, the Moriston, the Beauly, the Conon, the Shin and the other eastward-flowing rivers had all been mapped and surveyed by the end of the 1930s, and found ripe for exploitation. In 1921 the Water Power Resources Committee, chaired by Sir John Snell, estimated that the water power resources of Scotland, mostly in the Highlands, were capable of generating 1,880 million units per annum, or 217,965 kW. Another government committee, the Scottish Economic Committee (Hilleary Committee), examined the Highlands in 1938 and found a potential output of 1,972 million units per year.[1] During the 1920s and the 1930s, young engineers out from the Central Belt on hillwalking trips noticed the potential of particular sites and filed away their knowledge for future use. In the eyes of the engineers the glens and straths nursed lochs begging to be dammed, cataracts asking to be harnessed. There would of course be obstacles to be overcome, human and natural, but the engineers were confident that, given enough men and material, they could do the job.

Geology presented problems. Most of the upland had a thin skin of soil and vegetation over unyielding, dense rock of tremendous age. The metamorphic schists and gneisses were also shot through in places with granite intrusions from long-dead volcanoes and faultlines where earth tremors were regularly detected; and the surface had been gouged by ice to leave deep U-shaped valleys, jumbled ridges of clay and boulders, and rounded corries. On top of that there was the climate, but this was ultimately a boon. From the point of view of the engineers, the mountain belt fronting the westerlies was blessed with a high rainfall. In fact, it was drenched, with hardly less than sixty inches a year anywhere and over 120 inches in some particularly sodden spots. There would be no shortage of water, as all the workforce remembered: 'You'd stand on the scheme and you'd look up the hill and you'd see black spears and then you would look for the nearest bloody shelter because that was the rain coming and when it hit you, my God, it was hard.'[2]

Neither was the temperature always very comfortable. Snow lay from October to May on the higher ground and frost could strike in any month. Men up in the Affric hills working in topcoats would make a seventeen-mile bus journey down to Beauly and find the locals in their shirtsleeves, basking in a heatwave. And then, of course, there were the midgies, although the winds usually scurrying through the corries at least made them lie low most of the time.

This, then, is the story of the men who built the dams, tunnels, pipelines, power stations, and distribution lines; and of some of the consequences of their labours. I hope that I have not let any of them down in the telling and that they will forgive the errors.

Acknowledgements

In the course of researching the book, I interviewed many men and women who worked on the different parts of the schemes. They all gave their time willingly and, without their help, this project would have been impossible. Their accounts of their experiences in their own words are the heart of this narrative and bring the period and the endeavour vividly to life. I am grateful to them all for their hospitality and kindness. In alphabetical order they are: Henry 'Ben' Bentley, Dingwall; Ronald Birse, Edinburgh; Paddy Boyle, Glasgow; Archie Chisholm, Kirkhill; Bill Cooper, Inverness; Sybil Davidson and Mairi Stewart, Muir of Ord; Laurie Donald, Inverness; Pat Kennedy, East Kilbride; Alastair Kirk, Inverness; Patrick McBride, Donegal; Hugh McCorriston, Cannich; Barry McDermott, Clydebank; Jimmy Macdonald, Invergordon; Patrick McGinley, Donegal; Roy Macintyre, Gairloch; Bill MacKenzie, Muir of Ord; Hamish Mackinven, Edinburgh; Dougie Maclennan, Dingwall; Donald and Jean MacLeod, Golspie; Iain Macmaster, Acharacle; Sandy MacPherson, Inveruglas; Iain MacRae, Inchmore; Wodek Majewski, Barbaraville; Sir Duncan Michael; Stanley Mills, Auchlochan; John Farquhar Munro MSP; Roy Osborne, Ullapool; Paddy Paterson, Fort Augustus; Sandy Payne, Wester Clunes; George Rennie, Connel; William Rosie, John o'Groats; Hamish Ross, Dingwall; Ian Sim, Polmont; Bob Sim, Inverness; Otton Stainke, Kildary; Don Smith, Fortrose; and Don West, Cabrich.

Antoin MacGabhann, Letterkenny, Donegal, kindly sent me a letter describing his experiences as a student worker. Patrick Campbell, a native of Dungloe now living in New Jersey, sent me his published memoir *Tunnel Tigers*; and Moira McNicol in the Stirling Council libraries service supplied me with a copy of Gillean Ford's compilation of memories of the Breadalbane schemes. James Stevenson kindly photocopied for me his account of the construction of the Loch Sloy dam; and Tom Leith in Strathaven went to considerable trouble to get me a copy of Edward MacColl's 1946 paper on the schemes. Professor Peter Payne, whose own book remains the definitive official account of the achievement of the Hydro-

Electric Board and without which my task would have been much more difficult, was supportive of this work.

Dr Walther Bindemann in Edinburgh, Gisela Cumming in Inverness, Heinz Ohff in Berlin, J. Anthony Hellen in Tyne and Wear, and the staff of the German Consulate, Edinburgh, all tried to help me track down any former-POWs who worked on the schemes. Our joint efforts were unsuccessful but I am grateful to everyone for their cooperation freely given.

Jennifer Paice, Julian Reeves, Peter Donaldson and Heather Ward of Scottish and Southern Energy were immensely helpful in providing access to the Hydro-Electric Board archives. Angela Greig at Scottish Power likewise dealt with my queries with unfailing kindness. I would also like to thank Elaine Rodger and the staff at the Ben Cruachan Visitor Centre for their hospitality. Martin Broome, Roger Reid, Jim Donaldson and Louise Jones of Miller Civil Engineering Services Ltd welcomed me to the site of the Cuileig River scheme.

Many friends rallied to the task of helping me with accommodation, information or contacts, and in particular I would like to thank Roger and Anne Boulter, Dublin; Dr Jim Calder; Donald and Jean Campbell, Edinburgh; Isabel and Jimmy Gunn, Muir of Ord; and Alistair MacEachran, Raddery. I also owe a thank you to the editors of the *Derry People and Donegal News*, the *Donegal Democrat*, and the *Galway Advertiser*.

Research inevitably involves much use of libraries and I would particularly like to thank Paul Adair, Perth Museum; Perth Library, Local Studies Section; Highlands and Islands Enterprise; Inverness Library Reference Section; Rachel Chisholm, Highland Folk Museum; Carol Morgan, the archivist of the Institution of Civil Engineers; and Fiona MacCallum, Oban Public Library. Keith Moore, the senior librarian and archivist of the Institution of Mechanical Engineers kindly invited me to use their library but unfortunately I was unable to visit it.

Bob Sim, Archie Chisholm, Alastair Kirk and Hamish Mackinven read parts of the text and corrected several of my errors. Any that remain are all my own work. And once again I would like to thank Dick Raynor for his expert assistance with the photographs, my long-suffering agent Duncan McAra for his continuing support, and Hugh Andrew at Birlinn for his patience. Work on this book began when the author held a Hawthornden Fellowship.

INVERNESS
FEBRUARY 2002

Picture credits

The sources of the photographs are indicated in the individual captions, but I would like to record a special thankyou to Ann Yule; Donald Macleod; Ian Macleod, Dingwall Museum; Don West; Sybil Davidson; Archie Chisholm; Ronald Birse for the Report of the Mitchell Construction Company on the Moriston Scheme, and the copy of the *Civil Engineering and Public Works Review* with the article about the Breadalbane Scheme; Wodek Majewski; the *Aberdeen Press and Journal*; Nicholas McCormick of Edmund Nuttall Ltd; Hamish Ross; Iain Macmaster; Barry McDermot; Heather Ward of Scottish and Southern Energy plc for all the plates from the archive of the North of Scotland Hydro-Electric Board (NOSHEB) and for permission to adapt Hydro-Board maps to show the various schemes; Perth Museum and Louis Flood; and Carol Morgan, the archivist of the Institution of Civil Engineers, for permission to reproduce the four figures from the Institution's *Proceedings*.

Every reasonable effort was made to track down copyright holders and obtain permission for use of the illustrations, but if anyone has been overlooked, I hope they will accept this apology.

[1] '... the most hopeful thing ...'

Vast in size but thinly populated, the Highlands evoked opposing views in all who were concerned in the 1930s for their future. For many they were, in a phrase that came later, 'the last great wilderness in Europe', some 16,000 square miles of magnificent mountains, sprawling moors, mysterious glens and a wealth of wildlife that included the red deer, the golden eagle and the wildcat. For others the landscape represented a man-made wilderness, the sad result of decades of oppressive landlordism, evictions and social deprivation from which the only escape had been and still was emigration. Between 1921 and 1951 the population of the Highlands and Islands fell by around 15 per cent, from 371,372 to 316,471.[1] The land was being emptied of its inhabitants, and what to do to reverse this trend was the subject of many books, articles and reports, often peppered with such loaded phrases as 'the Highland problem' or 'the Highland question'.

Life in the Highlands had never been easy – the thin soil and the harsh winters saw to that – but surely something could be done. The Highlanders were an enterprising, intelligent people; they had proved their abilities time and again in every corner of the Empire, but somehow on their home ground they remained acquiescent and, the occasional land raid apart, not nearly as troublesome to politicians as their urban relatives.

'What is then at issue is not so much restoration of a prosperity which never really existed as the application of modern methods and modern knowledge to the old agricultural economy of the Highlands,' wrote Hugh Quigley.[2] Looking north from his suburban home in Esher, Quigley spoke for many who loved the Highlands but recognised that 'resurrection' (his term) was desperately needed. Tourism, forestry, fisheries, improved transport and the development of cottage industries were among the favoured options. The Forestry Commission, established in 1919, had planted thousands of acres with conifers in Argyll and the Great Glen, where Neil Gunn saw them in 1937 and considered their green spires: 'What he [the Highlander] wants now – where the spirit has been left in him to want anything constructive – is hope for the future, and these new forests along the banks of the Canal and on both sides of Loch Lochy were somehow like a symbol of a new order. The trees were full of sap, of young life, green and eager, larches and other pines, pointed in aspiration, and with an air about them not of privilege but of freedom.'[3]

The Second World War brought men and women once again from the glens to serve the country, and added another set of names to the memorials in every parish, but it also gave impetus to a sense that something had to be done and to a feeling that from the all-consuming effort of war would emerge a new future.

Industry in the Highlands had always been small and local in scale. Some processing of primary produce – the turning of grain into whisky and wool into tweed, the curing of fish – was established and significant; but the Highlands had no coal, apart from isolated mines at Brora and Machrihanish, and it was accepted that large-scale manufacturing belonged elsewhere, in the lowland cities where the labour force, markets and infrastructure favoured a concentration of effort. In the last decade of the nineteenth century, however, the potential of the region for water power had been realised. The North British Aluminium Company, formed in 1894, looked to the Highlands for a reliable supply of electricity, essential in the relatively new technology of converting raw bauxite to aluminium, and found it at Foyers on the south side of Loch Ness. Up to 19,000 kW of electricity were needed to convert four tons of bauxite to one ton of pure metal. Construction of the first major hydro-electric scheme in Britain began in 1895, and the smelting plant produced its first metal the following year, some 200 tons but already 10 per cent of the world output at that time. By 1900 production at Foyers had risen to over 1,000 tons, as the world demand for aluminium rose.[4]

Scotland's first hydro-electric plant for public supply had been installed at Greenock in 1885, only four years after the first in Britain opened in Godalming, Surrey.[5] The Greenock experiment ran for only two years but it had been enough to show the potential of hydro-electricity as a clean source of energy for daily activities. The next place to benefit from hydro-electric power was the village of Fort Augustus; in 1890 the Benedictine monks installed an 18-kilowatt turbine in one of the burns supplying their abbey at the southern end of Loch Ness and distributed the excess energy to their secular neighbours. The hotels and houses of the village were to have the benefit of this local supply until nationalisation of the industry in 1948. In

1896, the Fort William Electric Light Company began to operate two turbines at Blarmachfoldach on the Kiachnish River to supply light to the town. Another local scheme, this time at Ravens Rock in Glen Sgathaich, to the north of Strathpeffer, was built in 1903 with funding from Colonel E. W. Blunt-Mackenzie, husband of the Countess of Cromarty, and brought power to Dingwall and Strathpeffer. This enterprise was later transferred to a larger power station at the Falls of Conon on Loch Luichart. The coming of the new source of light was a wonder of the age. 'On Monday evening', reported the *North Star* in Dingwall, 'the electric light was turned on in the premises of Baillie Frew, jeweller, by his niece, Miss Christine Frew. The glitter and dazzle of the jewellery, caused by the numerous arc lamps, attracted great attention.' An ironmonger's and a bookseller's shop were also illuminated.[6] Blair Atholl received its first hydro-electricity supply in a similar way in 1910 when the Duke of Atholl built a 130-kilowatt generator on the Banrie Burn, a tributary of the Tilt, to supply his castle and the adjoining village.[7] Beyond the ends of the wires strung in these isolated localities, the people still depended on the oil lamp and the kitchen range and, in the countryside, were to do so for around another fifty years. These small beginnings had, however, been literally a glimmer of the future.

The aluminium industry continued to grow. A village grew up at Foyers to house the staff of the plant beside Loch Ness. The British Aluminium Company decided to expand its facilities and initiated an extensive scheme in the Loch Leven area that was to create the industrial village of Kinlochleven, with its smelting plant and the associated hydro-electric works drawing on the abundant water of Rannoch Moor. A dam was built across the Blackwater River to turn it into an eight-mile-long reservoir whose waters were then led down the mountainside to a power station above Kinlochleven. Construction began in 1905 and was complete four years later. As a major undertaking in remote mountain country, with the creation of a new loch and the redirection of existing water courses, it was a forerunner of what was to come.

It also marked the end of a more primitive era: the Blackwater Dam, 3,000 feet long and 90 feet high, in its time the largest in Europe, was the last large construction project built by the hard labour, unassisted by machinery, of itinerant Irish navvies. What that was like was captured in Patrick MacGill's novel, *Children of the Dead End,* first published in 1914 and based on the author's own experience of wielding shovel and drilling hammer in the uninhabited, waterlogged wastes of Rannoch Moor. The Kinlochleven project also attracted a large number of labourers from the Hebrides, so many in fact that foremen or 'gangers' had to have a command of Gaelic.

The navvies lived in shacks with tarred canvas roofs and slept in bunks, sometimes shared by three men, arranged in tiers around the flimsy walls. Cooking was done in frying pans on a stove in the centre of the muddy floor, and light was provided by naptha-burning lamps. There was almost no law and order among the 3,000 workers beyond what men could exert with their fists, and the only diversions were drinking and gambling. It was less a life than an existence. The highest paid workers, the hammermen, earned sixpence an hour, with rises to sevenpence-ha'penny for overtime and ninepence on Sundays. Drilling the rock was done by teams of five: one man, the holder, sat gripping the steel drill between his knees while his four companions struck it in rotation with sledgehammers until they had driven a hole four or five feet deep. Dynamite was then packed in the hole and the rock blown apart. 'We spoke of waterworks', wrote MacGill, 'but only the contractors knew what the work was intended for. We did not know, and we did not care.' MacGill also recorded how life in the camp rolled relentlesly and violently on without contact with the native Highlanders: the navvies were 'outcasts ... despised ... rejected ... forgotten'. A small graveyard with cement tombstones lies on a hillock a little to the west of the dam, the last resting place of some twenty of the navvies. The work camps associated with the later hydro-electric schemes had their share of violence, drinking and gambling but they were a world away from what MacGill and his mates endured.[8]

The First World War brought about a massive rise in the demand for aluminium and the Blackwater Reservoir had to be expanded to cope with the extra electricity requirement. Five hundred British troops and 1,200 German prisoners of war were brought in to build a five-mile aqueduct to lead water from Loch Eilde Mhor into the Blackwater. The British Aluminium Company set in train another development in 1924. Called the Lochaber project, it continued until the end of 1943. The main elements of this scheme were a 900-foot dam to divert water from the upper reaches of the Spey into Loch Laggan which, in turn, fed water through a tunnel to Loch Treig. A fifteen-foot diameter pressure tunnel was driven fifteen miles under the Ben Nevis massif to emerge at the head of a steel pipeline 600 feet above a power station in Fort William. The original plan to build an extra power station at Kinlochleven had to be shelved when Inverness County Council, in whose territory lay the Spey and the Laggan, refused to allow its resources to be piped across the county boundary to Kinlochleven in Argyllshire.

There were several schemes in the 1920s and 1930s to generate power for public use. The Clyde Valley Company's power stations on the Falls of

Clyde opened in 1926. The chief technical engineer on this scheme was Edward MacColl who was later to bring his expertise to the Hydro Board. A larger scheme in Galloway was built between 1931 and 1936. In the Highlands, the main effort was made by the Grampian Electricity Supply Company (acquired by the Scottish Power Company Ltd in 1927) and involved tapping Lochs Ericht, Rannoch and Tummel, with extra feed from Lochs Seilich and Garry, to generate electricity to serve a wide area of the central, southern Highlands and the Central Belt. The power stations opened in 1930 and 1933. The hydro-electric schemes of the interwar years established the pattern that was to be followed after 1945. They all employed large numbers of men – for example, 3,000 at the height of the Lochaber project – who lived in work camps and used technology to allow them to build and drill in the harsh landscape. Compressed air drills were deployed on boring out the pressure tunnel under Ben Nevis, and the workers had electrical power from a temporary generating station on the River Spean.

The Grampian scheme showed how Highland water could be harnessed for the public good and the Cooper Committee, sitting during the early years of the Second World War, looked with approval on its achievement. Not everybody was happy about the ambitions of the Grampian company and when, in 1929, they first put forward plans to develop the waters of the river system that discharged through the Beauly River into the Beauly Firth they met with considerable opposition. This plan would have involved the lochs of Affric, Mullardoch and Monar but it was rejected by the House of Lords, after strong arguments from A. M. MacEwan, the Provost of Inverness, and the Mining Association. Their combined opposition was based on the destruction of the beauty of this area of the Highlands and the fact that there were not enough consumers to benefit from the power to be generated.

Inverness had considered in 1921 accepting an extension of the power output from Loch Luichart to supply the town but the costs, estimated to be in the region of £230,000, made them cautious.[9] At that time there was not considered to be enough of a demand for electricity to justify the expenditure. A few years later the Town Council plumped for a turbine and generator installed in the Caledonian Canal on the southern outskirts of the town and, in 1926, this municipal initiative came on stream so successfully that in its first ten months it made a net profit of £7,000[10] and enabled the steam-powered generating plant in the town centre to be closed down at certain periods.

Throughout the interwar years private companies supplied electricity to several areas. For example, the Ross-shire Electric Supply Company, the firm founded in 1903 by Colonel Blunt-Mackenzie, had a transmission line running north up the Moray Firth seaboard from its generating station at Loch Luichart in Strathconon through Dingwall and the Easter Ross towns as far as Dornoch in Sutherland, where there was a switch-on ceremony in March 1933. John Murray, the Provost of the small Royal burgh, presided at the ceremony to which, in view of the short notice, a large crowd had been summoned by the town crier and his bell. The Provost's wife pressed the button to switch on six lamps: these shone with a 'cheery, mellow light while ... the street lamps shone forth in all their brilliance'.[11] A steam power station had been established in Perth in 1901 and was taken over by the Grampian company in 1933. Other small firms ran local generating plants in such towns as Crieff and Dunblane. In the south-west Highlands, there were private or municipal supplies in Campbeltown, Ardrishaig, Dunoon, Oban, Tobermory and a few other centres. A small plant had been installed at Gorten to supply Acharacle and Salen on the Ardnamurchan peninsula by K. M. Clark, the landowner, in 1928; this system remained in private hands until the mid-1950s when, in a sadly deteriorating condition, it was taken over by the Hydro Board.

Each proposed hydro-electric scheme had to receive parliamentary approval and it was during the lengthy process of consideration at Westminster that opponents could deliver the fatal thrust to kill a scheme dead. There were strong interests against hydro-electric power. The coal industry, Highland landowners and sportsmen made an unlikely but effective alliance against hydro-electricity. Some MPs and local authorities also voiced their objections, often basing their opposition on the perception of the schemes as simply another way in which Highland resources were to be exploited by lowlanders. If a private firm received the go-ahead for such a scheme, stated the *Inverness Courier,* the town would be deprived of 'valuable rights which are legally and morally hers' and referred to the support for the schemes from Fort William Town Council and the Lochaber Labour Party as 'base treachery'.[12] Some MPs argued for the schemes, acknowledging the growing importance of tourism and the need to conserve the landscape but also recognising that the Highlanders needed some industry to provide employment. In April 1938, when the Caledonian Power scheme, the latest proposal to develop the water power resources in Glen Affric, failed to survive the Second Reading in the House of Commons, the *Inverness Courier* printed a triumphant editorial: 'The opponents of [the Bill] have been falsely represented as being opposed to the development of water power and the introduction of industry in every shape and form. Nothing could be further from the truth. What we ... maintain is that there shall be no further development of the water power resources of the Highlands until a Committee is set up by the Government to enquire into [how] ... these water resources should be developed for the benefit of the Highlands.'[13]

The leader writer was presumably the *Courier*'s editor, Dr Evan Barron, whom we shall meet again. He could not have foreseen that in just a few years such a Committee would get down to business, thanks to the foresight and drive of one man.

Tom Johnston was both a socialist and an unrepentantly patriotic Scot. On the ship taking him and other British journalists to Russia for a tour of the Soviet state in 1934, he wore a Kilmarnock bonnet to declare, as he put it, his 'national status' (although he resisted appeals to do the Highland Fling). Born in Kirkintilloch in 1881, Johnston joined a cousin's printing and journalism business in Glasgow and launched the socialist weekly *Forward* in 1906. At the same time he cut his political teeth in local government, implementing innovative projects in adult education and municipal finance. In 1909, he published a scorching attack on the aristocracy in *Our Scots Noble Families*, a title dripping in irony as Johnston aimed to show how the prominent landowning dynasties had reached their eminent positions through robbery and fraud. The book was to prove to be an embarrassment to him later,[14] but it established him on the political stage and by the end of the First World War he was a leading figure in Labour politics. In 1922 he was elected as the Independent Labour Member of Parliament for West Stirlingshire but he lost the seat within two years when Ramsay MacDonald's Labour government fell to the Conservatives. A by-election a few weeks later brought Johnston back to the House of Commons; and in Ramsay MacDonald's second Labour administration between 1929 and 1931, Johnston briefly held Cabinet rank. He was returned to Westminster again in 1935.

On the outbreak of the War in 1939, Johnston was appointed Regional [*sic*] Commissioner for Civil Defence in Scotland. Then, in February 1941, Winston Churchill summoned him to Downing Street. The Prime Minister had already tried to persuade the craggy Johnston to accept a London post but now he had another plan. Johnston compared an interview with Churchill to being like a rabbit before a boa constrictor. When the Scot said he wanted to get out of politics to write history, Churchill gave a disdainful snort and said Johnston should join him and 'help ... make history'. The Prime Minister then laid his cards on the table: he wanted Johnston to be Secretary of State for Scotland. If Johnston felt himself to be like a rabbit, he remained a canny rabbit and agreed to take the post on certain conditions. The most important of these was that he could try out a Council of State comprising all five surviving former Secretaries of State and that whenever they agreed on a Scottish issue Johnston could look to Churchill for backing.

'I'll look sympathetically upon anything about which Scotland is unanimous,' Johnston records the Prime Minister as saying. 'What next?'

Johnston said he wanted no payment for the job as long as the War lasted. 'Right!' agreed Churchill. 'Nobody can prevent you taking nothing.'

Johnston said later that he was 'bundled out, a little bewildered', and miserable at the thought of the commuting he would have to endure between London and his beloved homeland; but he was also pleased that he had been given a unique opportunity 'to inaugurate some large-scale reforms ... which ... might mean Scotia Resurgent'. As he strode down Whitehall he was already listing the projects he was itching to start, and they included 'a jolly good try at a public corporation on a non-profit basis to harness Highland water power for electricity'.[15]

No one can be sure of the reasons for Churchill's choice of Johnston as Secretary of State. It is tempting to speculate that it may have stemmed from the Prime Minister's memory of the First World War and the Red Clydesiders but there is no evidence for this. Although he had never shared all the extreme left-wing views of some of the Red Clydesiders, Johnston had been a leading radical voice in that troubled time but had become more moderate since election to Westminster. Churchill may have been attracted by this poacher-turned-gamekeeper side to Johnston's career but he would also have known that Johnston was a highly respected man north of the border and was well equipped to keep the home fires loyally burning.[16]

Johnston's Council of State was officially named the Scottish Advisory Council of ex-Secretaries. The other members were Lord Alness, Sir Archibald Sinclair, Sir John Colville (later Lord Clydesmuir), Walter Elliot and Ernest Brown, and, by Johnston's account, they got on well, despite representing widely varying points on the political spectrum, and proposed projects and reforms in quick succession that laid the basis for postwar reconstruction in Scotland in a broad sweep of public life.

In 1938 Johnston had voted against the Caledonian Power scheme, sharing the opinion of many Highlanders that a private firm should not be allowed to take over a national resource. A Grampian company scheme for Glen Affric was again voted down in the House of Commons in September 1941; at the same time Johnston announced that the government had its own plans in train.[17] Johnston's view of hydro-electricity, in keeping with his socialist principles, was that public resources should be handled by publicly owned corporations. He had been impressed by the work of the Tennessee Valley Authority in the United States. The TVA, a government agency with the flexibility of a private corporation, was one of the most innovative ideas to emerge from the desk of President Franklin D. Roosevelt in his New Deal aimed to raise the American economy from the depths of the Depression. In the early 1930s, the valley of the Tennessee River was suffering severely from soil erosion and falling fertility, impoverishing thousands of farmers along its banks. As part of an integrated approach to the restoration of the area, the TVA built hydro-electric dams to provide power and control flooding, and integrated power generation into the rural landscape. It was an attractive model for what could be done in the Highlands.

The Cooper Committee was appointed in October 1941 to consider anew the potential for hydro-electricity generation in the Highlands. This body's official name was the Committee on Hydro-Electric Development in Scotland but it quickly became known by the name of its chairman, Baron Cooper of Culross. Thomas Mackay Cooper, son of an Edinburgh burgh engineer and a Caithness mother, had risen high in public service since graduating in law from Edinburgh University: he had been awarded the OBE in 1920 for his work in the War Trade Department, won the West Edinburgh parliamentary seat as a Tory in 1935, became a judge in June 1941 and was now Lord Justice General of Scotland. He was a firm supporter of Scots law and in one of his legal judgments had questioned the sovereignty of Westminster in relation to the Treaty of Union; this patriotic streak, as well as his intellect and his capacity for hard work, probably appealed to Tom Johnston.[18]

The other members of the Cooper Committee were the Viscount, William Douglas Weir, whose family background encompassed an engineering firm in Glasgow and who had served on the committee that devised Britain's national grid in the 1920s; Neil Beaton, the chairman of the Scottish Co-operative Wholesale Society and the son of a Sutherland shepherd; James Williamson, the chief civil engineer with the consultants to the construction of the Galloway hydro-electric scheme in the 1930s; and John A. Cameron of the Land Court.

Although 'handicapped by war conditions', the Committee examined every aspect of its remit throughout the first half of 1942. It combed through existing data, records and reports (the Snell Committee at the end of the First World War and the Hilleary Committee in the late 1930s had already considered hydro-electricity development in Britain). It also consulted the Electricity Commission, the Central Electricity Board, local authorities, power companies, industry, fishery boards, estate owners and representatives of a wide range of miscellaneous bodies, including the Royal Scottish Automobile Club and the Saltire Society. At the beginning Lord Cooper was sceptical of the Committee's ability to come up with much to supersede earlier work but, as the data accumulated, he became an increasingly enthusiastic supporter of hydro-electricity.

On its peregrination around the country, the Committee met Evan Barron, the editor of the *Inverness Courier*, in his office in the newspaper building on the east bank of the Ness; and Barron impressed on them the need for a *quid pro quo* if Highland water were to be harnessed.[19] In their final report, the Cooper Committee recognised that the 'portion of the area popularly designated the Highlands has for long been a depressed area and will remain so unless vigorous and farsighted remedial action is taken in hand without delay'. The Committee looked long and hard at not only the potential for hydro-electric development in the Highlands but also at some of the likely results of such development. To those who objected on what was then called 'amenity grounds', the Committee retorted sharply that 'If it is desired to preserve the natural features of the Highlands unchanged in all time coming for the benefit of those holiday makers who wish to contemplate them in their natural state during the comparatively brief season imposed by the climatic conditions, then the logical outcome ... would be to convert the greater part ... into a national park and to sterilise it in perpetuity, providing a few "reservations" in which the dwindling remnants of the native population could for a time ... reside until they eventually became extinct'.

'We accordingly recommend', wrote Lord Cooper and his colleagues, 'that there should be created a new public service corporation called the North Scotland Hydro-Electric Board' to be responsible for the generation, transmission and supply of power in all the parts of the Highlands currently outside the 'limits of existing undertakers'. (In time, with the nationalisation of electricity in 1948, the whole of the Highlands and Islands came under the aegis of the Board.)

As the Cooper Committee had been gathering its evidence, Tom Johnston had asked Evan Barron to come down to St Andrew's House to talk about how hydro-electricity might be made acceptable in the Highlands. Barron was ill at the time and sent a member of his staff with a written summary of his views. Such was the weight given to Barron's opinion that his submission was included in the final Bill almost word for word. Although the *Courier* editor was politically much further to the right than the Secretary of State, the two men respected each other highly and maintained a close friendship.[20]

Barron repeated his assertion that Highlanders would agree to hydro-electricity development only if they were to benefit directly from this surrender of their resources. Johnston responded by passing to Barron a draft copy of the Cooper Committee's report with the warning not to publish it:

Johnston is reported to have said, 'If anything appears about the Cooper report before Parliament gets it, Scotland will have another secretary of state next week.'[21]

Barron said nothing publicly until, three days after Johnston had laid the Cooper Report before the House of Commons on 15 December 1942, he was able to welcome it in an editorial.[22] In a long leader in the issue of the *Courier* on 23 January 1943, he broke from his usual comments on the progress of the War to give his opinion of the Hydro-Electric Development (Scotland) Bill under the headline, 'Hope for the Highlands'. The introduction of the Bill based on the Cooper Committee findings, he wrote, 'is the most hopeful thing for the Highlands which has happened for many a day'. Although the Cooper Report had included many 'mis-statements and misconceptions' it was the final recommendations included in the Bill that mattered and these, in Barron's opinion, conceded 'practically all we had fought for for twenty-five years'. The water resources of the Highlands were to be developed in the interests of the native Highlander. Barron dismissed the objections raised in editorials of other newspapers such as *The Scotsman*, which he saw as being the voice of Big Business, and called on Highlanders to see that the Bill became law more or less as it stood and put their water resources 'forever beyond the reach of the clutching hands' of outside companies. Now, said the *Courier*, the State had the chance to undo the ill-treatment meted out to the region for the last 150 years and, reminding the Secretary of State of the service to the nation being rendered by the 51st Division, at that time slogging through the African desert, declared the passing of the Bill was not a favour but the fulfilment of a duty.[23]

The Bill recognised the broader role of the Board in what became known as the social clause: this stated that the profit from the sale of surplus electricity to the Central Electricity Board for the national grid would be ploughed back into reducing the costs of distributing power to the more remote, low-populated areas of the Highlands for 'the economic development and social improvement' of the region. In May 1943 the *Courier* was pleased to say that Tom Johnston had 'earned the gratitude of all who love the Highlands and who believe they have a future as great and as noble as their past'.[24] After collecting a few minor amendments on its passage through Parliament, the Hydro-Electric Development (Scotland) Act became law in August. Writing of the opponents of the Act, Johnston returned to the rhetoric of his younger days:

> I knew most of the nests from which the corbies would operate; the colliery owners had retired from the struggle, and their share-

> holders wanted no notice taken of the pit bings and so stopped talking about how the hydro schemes would destroy amenity. A few shameless twelfth of August shooting tourists, who themselves took care to live in the electrified south for eleven months in the year, moaned about the possible disappearance in the Highlands of the picturesque cruisie; and I had one deputation whose spokesman was sure we were engaged in a conspiracy to clear Glen Affric of its crofters and its sheep; in response to enquiries, he had not been up at Glen Affric himself, and he really was surprised to learn that there were neither crofters nor sheep in the Glen for these many years past.[25]

In September the names of the first members of the Board were made public. The Earl of Airlie was appointed chairman, with Edward MacColl as deputy chairman and chief executive. After his success on the Falls of Clyde scheme, MacColl, whose forebears came from Melfert in Argyllshire, had been appointed engineer for the Central Scotland District of the Central Electricity Board and had overseen the construction of the first regional grid in Britain. He brought a vast experience of the technical aspects of electricity generation and distribution to the Board, and added to this formidable expertise a flair for innovation. The other three members were Neil Beaton, who had already served on the Cooper Committee; Hugh Mackenzie, the Provost of Inverness; and Walter Whigham, a director of the Bank of England and the representative of the Central Electricity Board. (Whigham was soon to resign through ill health and his place was filled by Sir Duncan Watson, a Scottish engineer.)

The Earl of Airlie seems at first glance to have been an unlikely choice for the figurehead of a new public corporation. He was the twelfth member of his family to hold the Airlie title, had been educated at Eton, had won the Military Cross in The Black Watch during the First World War, owned around 40,000 acres, was Lord Lieutenant of Angus, a member of Angus County Council and a staff officer at Scottish Command HQ. The good-natured Airlie had, however, been Tom Johnston's second-in-command when he had been in charge of civil defence, and the two men obviously felt that they could work well together.

Two sub-committees of the Board were set up – the Amenity Committee under the chairmanship of Colonel the Hon Ian Campbell and including Lady MacGregor of MacGregor, the only woman in the upper echelons of the Board; and the Fisheries Committee with Colonel Sir D. W. Cameron of Lochiel in the chair. The registered office was established in Edinburgh, and the Lord Lyon King of Arms granted the Board its own coat of arms in 1944.

Plate 1.
The shield of the North of Scotland Hydro-Electric Board as depicted in wrought iron on the gates of Invergarry power station (*author*).

The shield bore a winged thunderbolt emitting forked flashes of lightning suspended above a cruisie-lamp, the ancient form of domestic illumination. These symbols, encapsulating the Board's aspirations, were supported by two rampant stags on either side of a fir tree and a rock from which water gushed. The motto was in Gaelic: *Neart nan Gleann*, the power of the glens.

The Board benefited in its early decades from the calibre of the almost-handpicked senior staff – 'men steeped in their subject', according to Hamish Mackinven. Edward MacColl selected Angus Fulton, as enthusiastically in favour of hydro-electric development as himself, as his chief civil and hydraulic engineer; and wooed David Fenton back to Scotland from the English Midlands to be his commercial engineer. Thomas Lawrie became the Board's secretary on its inception. W. Guthrie was appointed as the first chief electrical engineer and A. N. Ferrier as the chief accountant.[26]

Inverness Town Council organised a conference in August 1943 where representatives from all the Highland and Islands local authorities could discuss the implications of the new Act.[27] Fearing that once again Highland resources might be exploited for the benefit of others, the so-called Scottish Local Authorities Hydro-Electric General Committee that emerged from the conference resolved to 'watch the interests of the area'.[28] For example, John Murray, the Provost of Dornoch, while calling for a bold policy to take advantage of the new source of energy and expressing confidence that industry would follow power, was concerned that the remote places wouldn't be forgotten.[29]

In March 1944, the Board published its development programme and listed no less than 102 projects, ranging in size from small local ones to giant schemes covering whole series of glens. At one end of the spectrum lay the streams draining into Loch nan Gillean, near Plockton, calculated to be capable of generating four million units (kilowatt-hours per year), the streams on Islay and Jura (five million units), two streams on the north side of Loch Nevis (five million units), and streams in Arisaig (six million units). The biggest schemes pinpointed the Affric-Beauly river system (440 million units), the Orrin-Conon and the Garry-Moriston systems (each 350 million units), and the Tummel-Garry system (300 million units).[30] It seemed as if every corner of the Highlands and Islands were included, from the burns on Shetland to those draining the Mull of Kintyre. The impressively ambitious programme recorded a total potential output of 6,274 million units of electricity per year, considerably more than the 4,000 million units per year estimated by the Cooper Committee. Edward MacColl pointed out in an address to the Institution of Engineers and Shipbuilders in Scotland that the programme did not include 'a substantial amount of power still available in the form of high-head run-off schemes' with little or no storage capacity in the form of lochs. He also conceded that not all the schemes in the list of 102 were economic 'when compared with other means of producing power', although in the future when coal became scarce or dear they might become viable.

The prospects were, however, exciting enough. 'Just before the War finished, the Ministry of Information made a film to show how good it would be to have power in the Highlands,' said Archie Chisholm, who was a schoolboy in Strathglass at the time. 'I remember seeing the team coming to make part of the film. They had my grand-uncle, Jim Simpson, with a pair of Clydesdales and a horse plough ploughing up on a very barren bit of ground. This was supposed to show worthless ground that was to be recovered. We were all supposed to get power for nothing. The idea was that the people coming home from the War would get better things. Of course we had no power then; unless you lived on an estate where there was a wee water turbine or a generator, it was the Tilley lamp, double-wick lamps and candles.'

The Scottish branch of the Association of Scientific Workers, a body firmly in favour of centrally planned, publicly owned advancement, hailed the Board's development programme by issuing a brochure *Highland Power*, which made direct reference to the Tennessee Valley Authority ('one of the greatest sociological experiments of history') and stated that the proposed developments offered 'a golden opportunity to test a new approach to British social and economic problems'. Other bodies more concerned with what might result when a great concrete dam was thrown across a glen also soon made their voices heard. In the summer of 1944, the Association for the Preservation of Rural Scotland protested that areas of outstanding natural beauty, such as Glen Affric, Glen Garry and Loch Maree, should be safeguarded.[31]

The schemes in the counties of Perth, Dunbarton, Argyll and Inverness were already being surveyed and planned in the spring of 1944[32] and the Board published the details of its first construction projects on 3 July. There were three – Loch Morar, Lochalsh and Loch Sloy, costing a total of £4.6 million and aimed at generating an estimated 136,000 units of electricity. Two were mainly of local significance: the Morar scheme proposed a dam and power station on the Morar river to provide power to the Mallaig and Morar area; and the Lochalsh scheme comprised a dam on the Allt Gleann Udalain and a power station near Nostie Bridge to meet local power needs. The third scheme, the one at Loch Sloy, was by far the largest of the three, a major enterprise involving the construction of a power station on the shore of Loch Lomond, four miles north of Tarbet, to be fed from a dam at Loch Sloy in the hills overlooking the outfall.

As 1944 wore on, opposition to the Board's proposals grew louder and more public. Letters began to appear in newspapers. R. Gilmour probably spoke for many when he wrote from the Lochboisdale Hotel on South Uist to say that the Board's intentions should be made 'crystal clear' to the people and that industrial factories were not desirable in the Highlands and Islands.[33] At a public meeting in Pitlochry, a motion was passed to express 'grave concern' about the proposals to create dams on the Tummel and the Garry and drown parts of the river valley.[34] A committee was formed to oppose the Tummel-Garry scheme; the charge that the large landowners in Perthshire lay behind the opposition was rebutted.[35] A few spoke up in support of the development, pointing out that a new loch might enhance rather than destroy the scenery, and someone using the sobriquet 'Beauty Lover, Perth' wrote 'We want Scotland to be a place where we can get a job after the War'.[36] The Hydro-Electric Board Amenity Committee met the Pitlochry Amenities Preservation Committee in Fisher's Hotel in the town to hear the local objections in detail, while other gatherings of the objectors took place in bars and hotels around the county.[37] The owners of the salmon fisheries appealed in vain to Perth Town Council for financial support in their campaign.[38] Perth County Council received a report on the likely effects of the scheme: Loch Tummel was predicted to rise seventeen feet and submerge 770 acres, some buildings would go under, and the new loch behind the proposed dam at Pitlochry would drown 165 acres.[39]

The members of the Board took to the road on the public relations offensive. Edward MacColl protested that confidential Board information was being used by anti-hydro agitators.[40] In Pitlochry, Board member Neil Beaton declared that work on the schemes would employ 10,000–12,000 Scots in the construction phase, that the dams and power station would provide up to £30,000 per year to the Perthshire rates bill, that a permanent staff of fifty would work in Pitlochry, and that hydro-electric schemes had brought about a rise in tourism in the Tennessee Valley, Switzerland and other countries. Beaton also gave voice to a matter hanging in every mind in late 1944: '...What was the position of Scotland before the War? A large section of the people were unemployed ... Many of these were fighting and unfortunately dying that Scotland might continue to live in freedom. Were these brave men and women to come back to the old conditions ...?'[41] Lord Galloway, the chairman of the Association for the Preservation of Rural Scotland, countered that the prewar hydro projects of the Grampian Electricity Company at Rannoch and Tummel Bridge had employed mainly

Irish labour, that the rates bill would not counteract lost tourist revenue, and that the permanent jobs resulting from the Pitlochry works would not compensate the families whose land would be flooded.[42]

The Local Authorities Committee was also raising doubts about the schemes. Some of these focused on who should control local water resources. Dunbarton County Council opposed the Loch Sloy scheme, the first Board project, because it might need Loch Sloy for domestic water.[43] Inverness County Council lamented the lack of information available from the Board on its plans for their area: a dispute between Inverness Town Council and the Board over who should have control over Loch Duntelchaig, whether the waters of this relatively small loch a few miles south of the town should be part of a hydro-electric scheme or reserved for domestic use, was to run for many months before it was settled in favour of the Council just before Christmas 1945.[44]

In April 1945, the Board published its first Annual Report covering the period between its inception and December 1944, a modest eight-page document in a brown cover, priced 6d. It summarised the progress so far: the approval of the Development Scheme by the Electricity Commissioners and its confirmation by the Secretary of State; the collation of existing rainfall records – they were found to be inadequate; the establishing of automatic river flow recorders on the Tummel and the Conon, with observers taking manual readings on four more; the collection of geological data in relation to the siting of dams, power stations, tunnels and aqueducts; and the publication of the first constructional and distribution schemes. Surveys were proceeding on more distribution schemes as quickly as the wartime staff shortages would allow.

Under the heading 'Future Policy' the Report stated that all the schemes for the supply of ordinary consumers in the Board's district appeared to be uneconomic. 'In the aggregate, when they are carried out, the annual loss will be very large and it will have to be covered by profits earned in other directions.' This meant selling electricity to the areas of high demand in the south. 'Projects of the same exporting type as the Loch Sloy Scheme are required and are being prepared, which will harness the undeveloped resources of the Highlands and help to pay for the many uneconomic distribution schemes there, and to finance a "Grid" in the North of Scotland District.' The intention to develop economic schemes before working on uneconomic ones for Highland use aroused the ire of some Board-watchers. It was against the terms of the Act's social clause, in the opinion of the *Inverness Courier*,[45] which clearly stated the Highlands had first claim on power from the glens. Lord Airlie said power had to be made available to

Map 1.
Sloy/Shira

redevelop postwar Lowland industry but arguments that economic schemes had to have priority in order to subsidise the others cut little ice beyond the Highland Line and words such as 'betrayal' began to appear in the editorials of Highland newspapers.

In its first sixteen months, the Board's total expenditure amounted to almost £137,000. This was met by temporary loans from Scottish banks. All during its life, the Board never received a penny in subsidy from public funds, as Tom Johnston and many others always stressed, although under the terms of the 1943 Act the Treasury guaranteed the Board's borrowings.

Objections to the Sloy scheme, from local authorities and from individuals, forced the holding of a public inquiry which got down to work in Edinburgh between Christmas and Hogmanay in 1944, with John Cameron KC as chairman. To Cameron's regret, the business took as long as six days (some later inquiries were to take considerably longer) but the outcome was a victory for the Board. In recommending that the scheme go ahead, Cameron did, however, remind the Board that it should be better prepared to argue its case in future. Dunbarton County Council's objection that it might need Sloy for its own purposes was dismissed, along with other fears over the appearance of pipelines, the power station and spoil disposal.[46] As Secretary of State, Tom Johnston happily gave the scheme the green light and, after resting before Parliament for the statutory forty days, it became finally clear to proceed on 28 March 1945.

The strong feelings of opposition seem to have been assuaged somewhat by publication of the Board's distribution schemes for Highland areas. The Gairloch-Aultbea distribution scheme, made public in December 1945, promised lines running from the power station on the River Kerry to bring electricity to some 1,500 people in an area of 180 square miles around Loch Maree and Loch Ewe.[47] Consumers within a reasonable distance of the power lines would be given free connection. The charges for electricity were likely to be 5d per unit for lighting, three farthings for cooking and heating and a halfpenny for other uses. The distribution scheme for Skye, published in February 1946, aimed to serve over 10,000 people in an area of 690 square miles, from a power station on the Bearreraig river and from a submarine cable across the Kyle of Lochalsh.[48]

On 11 June 1945, Margaret Johnston, the Secretary of State's wife, inaugurated the Loch Sloy scheme by cutting the ceremonial first sod on the site of the temporary diesel power station by Loch Lomond. In keeping with the scale of the enterprise, the 'sod' was a strip of turf twelve feet wide and 100 feet long, and the 'spade' was an eighteen-ton bulldozer rejoicing under the name of 'Red Lichtie'.

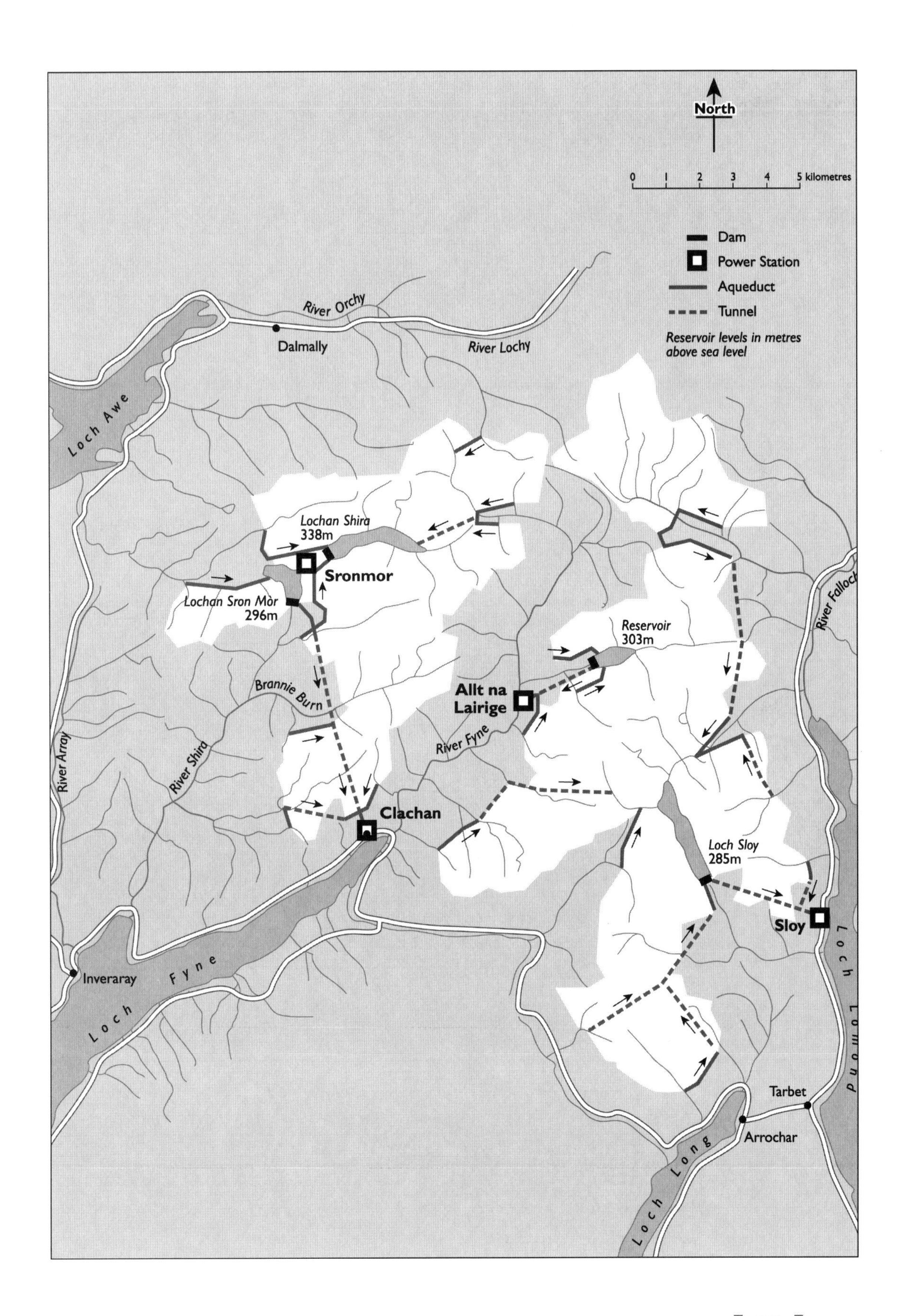
North
0 1 2 3 4 5 kilometres
Dam
Power Station
Aqueduct
Tunnel
Reservoir levels in metres above sea level
River Orchy
Dalmally
River Lochy
Loch Awe
Lochan Shira
338m
Sronmor
Lochan Sron Mòr
296m
Reservoir
303m
River Falloch
Brannie Burn
Allt na Lairige
River Fyne
River Array
River Shira
Clachan
Loch Sloy
285m
Sloy
Inveraray
Loch Fyne
Loch Lomond
Tarbet
Arrochar
Loch Long

Plate 2.
Tom Johnston in younger days (*Ann Yule*).

The crucial factor in the viability of a hydro-electric scheme is the amount of water available and the vertical distance, the 'head', through which it can be induced to fall to reach the power turbines. The hills to the west of Loch Lomond had plenty of water in the 1940s. Around Loch Sloy, rain fell on average on 230 days in every year, to a measured annual depth of some 120 inches. The loch had been surveyed in 1937 and the results of this earlier investigation were revived and modified by the Board engineers in 1944. Before the work began, Loch Sloy was a modest mile or so long, a shallow body of brown, peaty water nestling 780 feet above sea level between the bare rocky slopes of Ben Vane and Ben Vorlich. Very few people lived in the surrounding hills. At the east end of the loch, a burn gushed out through a gorge suitable for a dam and coursed down to swell the Uglas Burn on its steep descent to Inveruglas Bay and the waters of Loch Lomond some 600 feet below. To swell the size of the primary reservoir, a series of tunnels and aqueducts were constructed to divert water from other burns in the surrounding moorland. Loch Sloy, originally enjoying a catchment area of six and a half squares miles, finished up draining some twenty-seven square miles spread over the hills where the three counties of Dunbarton, Argyll and Perth met. The loch rose 155 feet in height and doubled in length, making its surface at its maximum over 900 feet above the turbines it was designed to feed.

Although the Loch Sloy scheme was fairly straightforward on paper, problems arose during construction and taught a few valuable lessons. It was planned to come on stream in 1947 but the difficulties of construction in the immediate postwar years caused major delays. There were acute shortages of almost everything – steel, cement, equipment, timber and men. The shortage of timber led to the use of steel shuttering. The weather remained atrocious – only three weeks without rain were to be recorded during the entire three years of dam construction – and the perpetual rain, sleet and driving winds sapped the will of many workers. The weekly average precipitation during 1947, 1948 and 1949 was 2.75 inches, and 'severe gales occurred with disheartening frequency'.[49] On top of this, the accommodation was rough, and the food became another deterrent. The Irish navvies at Kinlochleven in 1909 may have roughed it out, but this was the 1940s and expectations were higher. Lorries bringing materials took time to negotiate the single-track road winding along the shore of Loch Lomond.

Preliminary work took up most of the first two years of the scheme. An access road was driven up the steep valley of the Uglas to the slopes of Ben Vorlich, where part of it around the east side earned the nickname of 'the Burma road'. The camp and workshops had to be built, a bridge had to be

Plate 3.
Loch Sloy dam, January 1949 (*NOSHEB*).

Plate 4.
Loch Sloy dam, February 1949 (*NOSHEB*).

Plate 5.

Loch Sloy. Shuttering on the upper section of No. 9 buttress, October 1949 (*NOSHEB*).

Plate 6.
Loch Sloy dam, August 1949 (*NOSHEB*).

Plate 7.
Loch Sloy dam, October 1949 (*NOSHEB*).

put up to carry the West Highland railway line over the pipeline to the power station, a 3,600 kW diesel generator had to be installed to provide power for the construction. Portland cement was available only in bags, which took considerable handling. Sand had to be towed in barges up Loch Lomond from Balloch twenty-two miles away; and of course the sand had to be unloaded into five-ton lorries for the journey up the access road to the batching plant beside the dam. A conveyor-belt system was erected to carry crushed stone from a quarry on Ben Vane across 1.75 miles of rugged moor to the same batching plant. In the batching plant, the sand, cement and stone were mixed in careful quantities to produce the concrete that was then swung in ten-ton skips on an electrically driven overhead cableway to the site for pouring on the dam. All of this became standard procedure on future schemes.

The figures convey something of the scale of the work. The dam was not as large as some that would be constructed later but to make room for its foundations 36,000 cubic yards of peat and sandy soil, and 56,000 cubic yards of rock had to be scraped or blasted away. The dam itself, designed by James Williamson to be economical with materials, still had thirteen massive buttresses soaring rib-like from the floor of the gorge and consumed 208,000 cubic yards of concrete.

The main tunnel was driven from four entry faces. Drilling the fractured schist with compressed-air-powered bits was an arduous task. An advance of 64 feet became the weekly norm but in one glorious seven-day burst 103 feet of rock was carved out. The workers seem quickly to have acquired expertise for, when the four tunnels linked up, the error in alignment was less than one inch. The completed tunnel was ten thousand feet long with a maximum diameter of 15 feet 4 inches. To make this hole through Ben Vorlich's innards, the men had removed 180,000 tons of rock and fired off 220 tons of gelignite.

All the tunnels designed to carry water were equipped with surge shafts, vertical tunnels often with subsidiary expansion chambers, to accommodate sudden changes in water level as the load varied on the turbines in the power stations. The surge shaft at Loch Sloy was drilled out and lined with concrete to leave a space 26 feet in diameter and 273 feet high. The main tunnel led from the dam to the valve house on the edge of the brae overlooking Loch Lomond. Here the water flow was channelled into four steel pipelines that dropped steeply down the hillside to the power station. Built *in situ* by Sir William Arrol and Co., the pipeline, 3,500 tons in total weight, was designed to accommodate a flow of up to 220,000 tons of water per hour. The steel was accordingly graded in thickness and carefully welded in Arrol's

LOCH SLOY PROJECT
Contract 22 - Loch Sloy Dam
No. 17 Buttress Cut-
Ser. No. 79 Date: 13/9/49

Plate 8.
Opposite. Excavating the base of a buttress, 1949 (*NOSHEB*).

Plate 9.
Top. Loch Sloy dam nearing completion, August 1950 (*NOSHEB*).

Plate 10.
Bottom. Loch Sloy dam nearing completion, August 1950 (*NOSHEB*).

LOCH SLOY PROJECT
Contract 22 - Loch Sloy Dam
looking on Downstream North Face of Dam
Ser. No. 117 Date: 5/5/50

LOCH SLOY POWER HOUSE & PIPE LINE
General View
Date: 25/3/48

Plate 11.
Opposite top. Loch Sloy dam – final stages of the pouring of concrete on the dam wall, May 1950 (*NOSHEB*).

Plate 12.
Opposite bottom. Loch Sloy, March 1948. An early stage in the construction of the pipeline feeding the power station on the shore of Loch Lomond (*NOSHEB*).

Plate 13.
Loch Sloy, June 1949. Constructing the pipeline to the power station (*NOSHEB*).

workshops in Glasgow to make pipe sections to cope with pressures double the expected 400 pounds per square inch. The pipe sections were brought to the site by rail and hauled up the hillside to be laid on massive supporting blocks of concrete and concrete piers before being welded in place.

While all this was going on, the power station itself was being built by Hugh Leggat Ltd. The laying of the foundations for the turbines required the excavation of 37,000 tons of rock and earth. Progress was slow, bedevilled by the same problems that beset the other components of the scheme, but finally in 1948 the installation of the English Electric turbines went ahead. The four 32,500 kW generating sets, the most powerful so far deployed in Britain, were not all finally in place until nearly the end of 1951.

Electricity from Loch Sloy was destined to feed into the national grid at Windyhill on the fringe of Glasgow. The erection of the steel pylons, the towers, to carry the high voltage, 132 kV power lines across the countryside was almost as much a feat as the building of the dam and tunnel. Some vehicles fell victim to the peat bogs, and pack horses were brought in to help carry at least a little across the moors. Payne summarises the conditions in which the work was completed: 'Working from sunrise to sundown in incessant rain, paid about £8 for a seventy-hour week, with thirty shillings a

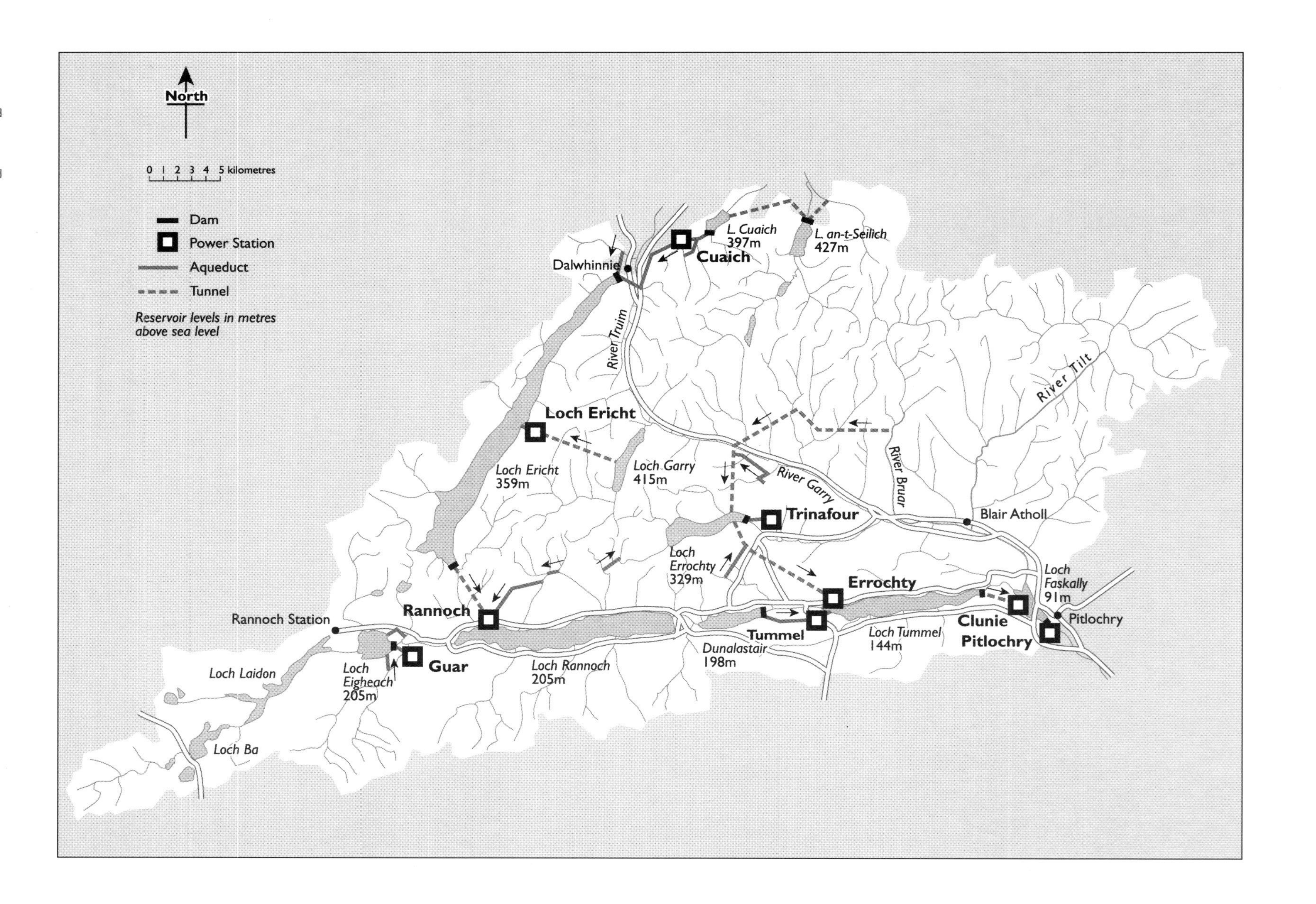
North
0 1 2 3 4 5 kilometres
Dam
Power Station
Aqueduct
Tunnel
Reservoir levels in metres above sea level
Dalwhinnie
Cuaich
L. Cuaich 397m
L. an-t-Seilich 427m
River Truim
River Tilt
Loch Ericht
Loch Ericht 359m
Loch Garry 415m
River Garry
River Bruar
Trinafour
Blair Atholl
Loch Errochty 329m
Errochty
Loch Faskally 91m
Pitlochry
Rannoch Station
Rannoch
Tummel
Loch Tummel 144m
Clunie
Pitlochry
Loch Laidon
Loch Eigheach 205m
Guar
Loch Rannoch 205m
Dunalastair 198m
Loch Ba

week lodging allowance, the labour force was exceptionally volatile: no less than 1,285 men were taken on during the course of the two-year contract to keep a squad of 200 going.'[50]

Map 2.
Tummel

As the Sloy scheme was slowly becoming a reality, the Board was facing a severe test to the east. Unlike the Sloy development, which was designed to supply Glasgow with electricity and which was being built in a relatively deserted spot that excited few passions, the Tummel-Garry scheme was sited in the heart of beautiful countryside with many historical associations. Hamish Mackinven considers opposition to the scheme to have been one of the three periods of greatest danger in the life of the Board. Some of the reasons for this have been mentioned already: Pitlochry feared for the damage to its tourist trade, large landowners did not want anything to threaten their interests in salmon fishing, and the slowly growing environmental lobby wanted to preserve the beautiful scenery in central Perthshire. Perth and Kinross County Council, led by the provost, G. T. McGlashan, expressed its unanimous opposition in March 1945 and criticised the Board for failure to keep the public informed about its intentions and to respond fully to the Council's repeated requests for information. The councillors included Lord Mansfield, who referred to the Board's 'miserable policy of secrecy', and Lord Kinnaird, who said the Council had been profoundly shocked to learn the Tummel-Garry scheme had been scheduled so soon in the Board's programme. In the *Perthshire Advertiser*, the editor said the Council was bound to oppose the scheme but the paper's columnist, Neil Johnson, took a dissenting view: writing that the Board offered reasonable hope for an end to the chequered economic history of the Gaels, he said it '... has to shoulder a stern business proposition ... shorn of that inane impractical romance which would seem to malign the judgement of those whose creed is the preservation of the beauty of the ... glens and straths at all costs'.[51]

The Tummel-Garry scheme was vital to the Board's programme. Designed for a capacity of 150,000 kW, the scheme was to provide energy principally to the Central Electricity Board but also to the Grampian Company and the city of Aberdeen, all sources of revenue that would enable the Board to proceed with smaller, loss-making schemes. The public inquiry opened in the august surroundings of Parliament House, Edinburgh, on 25 April 1945, with John Cameron KC once again in the chair. On this occasion he had Sir Robert Bryce Walker and Major G. H. M. Broun-Lindsay to assist him. A procession of advocates represented the twenty-five formal objectors whose complaints were almost all on amenity grounds and subjected Board witnesses to searching and at times hostile questioning. The same level of antagonism had also been found 'on the ground'. In Pitlochry only one hotel had been willing to offer accommodation to Board engineers when they were completing surveys. Lord Airlie was seen as a traitor to his class and his son was blackballed by the Perthshire Hunt.

Airlie lamented the parochial attitude of the objectors.[52] He was severely shaken by the tone of the cross-examination he had to undergo at the hands of some of the lawyers. Edward MacColl was unfortunately too ill at the time to attend the Inquiry. Other Board representatives were subjected to hostile questioning. Tom Lawrie, the secretary of the Board, was asked if visitors to Pitlochry would be coming 'to view your dam or to damn your view'. Lawrie said they would come anyway and reminded the Inquiry that the Tennessee Valley Authority schemes attracted two million tourists a year.[53] An advocate speaking on behalf of Atholl Properties Ltd regretted that the scenic beauty of Pitlochry would be converted into cash to provide electricity for Orkney.[54] Another, acting for the National Trust for Scotland, added two cutting lines to a well-known Jacobite song:

> Cam ye by Atholl lad wi the philabeg,
> Doon by the Tummel and banks o' the Garry,
> Saw ye the lads wi their bonnets and white cockades
> Leaving their land to follow Prince Cherlie.
>
> Saw ye the lads wi their cusecs and kilowatts
> leaving the rivers defaced by Lord Airlie.[55]

The Board stuck grimly to its guns and argued constantly that their plans for the Tummel-Garry were being carried out in the national public interest. The Inquiry dragged to a close just as VE Day was being celebrated and finally it was victory for the Board as well. Approval of the scheme finally came through towards the end of August when the Secretary of State released the order for the work to start.[56] Rumblings of opposition continued: Perth and Kinross County Council confirmed their resentment of it by twenty-eight votes to eleven in October 1945 and a motion was brought before Parliament a month later to annul the Secretary of State's order. The latter was defeated, signalling it was now too late for the objectors to win the day.

The Board had bought the Fonab estate, where the dam was to be, and German prisoners of war had been drafted in to widen the roads and do some of the preliminary construction work.[57] All the activity seemed to jerk

Pitlochry into ambition: it applied successfully for burgh status shortly afterwards[58] and gained a new twelve-acre recreation ground in 1948 at the Board's expense to replace the one about to go under water, just as work was beginning beside the Tummel beyond Clunie Bridge. In 1949, an article in *The Scots Magazine* noted that the 'more enterprising' of the village's hotel-keepers were drawing attention to their proximity to the hydro-electric scheme.[59]

The psychological bruising he had received during the Public Inquiry proved too much for Lord Airlie. He probably realised that he did not have the power to protect the Board from its enemies and see its development programme to fruition, and he resigned after Tom Johnston agreed to become the chairman. The change of command took place on 1 April 1946. Johnston was to remain as unpaid chairman of the Board until 1959, years during which the organisation he had brought into being grew to become a major feature of Highland life.

The Countess of Airlie was the guest of honour at the inauguration of the Tummel-Garry scheme on 25 April 1947. Some three hundred people braved the snow and heavy rain to see the ceremony. The Vale of Atholl Pipe Band marched. Tom Johnston regretted the delay in the start of the scheme, pointed out yet again the benefits that would ensue (including £160,000's worth of rates relief to Highland counties), and promised that the Board would do its best to get rid of the scars on the landscape. The Countess then set in motion a cement mixer to create the first foundation block for the new dam. A time capsule containing the front pages of that day's newspapers, coins ranging in value from a ha'penny to half a crown, the Airlie coat of arms, a copy of the Hydro-Electric Development Act, and a description of the Tummel-Garry scheme, was prepared for entombment in the concrete. Provost McGlashan recalled how the Council and the Board had not been seeing eye to eye two years before and how he could hardly have imagined he would be speaking this day in the enemy camp. But that was the way in British public life, he said; opponents shook hands after a severe fight and became better friends than before. 'We as a County Council have now established a most friendly relationship with the Board,' he concluded.[60] The *Perthshire Advertiser* had already noted in a leader that the Scottish Tourist Board did not fear damage to amenity, adding wryly that this was possibly because its president was none other than Tom Johnston, before going on to state a belief that 'even in their unbecoming infant shape' the dam and the power station 'will be an asset of tourism'.[61]

Hatchets buried in Perthshire, the Board moved on. Details of further construction schemes emerged apace from 16 Rothesay Terrace in the

Plate 14.
Grudie Bridge power station, June 1953 (*NOSHEB*).

succeeding twenty years. In February 1945, along with the Tummel-Garry scheme, the Fannich scheme, the first of the developments set to take place in the centre of Ross-shire (costing £6.45 million) was announced; and in September 1946 the Affric-Cannich scheme (£4.8 million), so contentious before the War when it had been the subject of a private initiative, was laid before the public. In this new incarnation, Loch Affric itself was to be left untouched and the main reservoir for the scheme was to be made at Loch Mullardoch in Glen Cannich to the north. From Mullardoch a tunnel would lead the water down to Loch Benevean at the east end of Glen Affric. Benevean would be dammed but its level would rise only about twenty-five feet and it would not change much in appearance. The Benevean water would then be fed through a tunnel to a power station at Fasnakyle on the floor of Strath Glass. Although several properties would be drowned in the enlarged Loch Mullardoch, only one house in Glen Affric would be submerged, and generally the scheme met with wide approval. The *Inverness Courier* noted that two thousand men would be employed on the construction and that the County Council coffers would receive some £10,000 in rates instead of the paltry £300 the glens were affording at present.[62]

Between 1945 and 1969, when work began at Foyers on the last of the 'big' hydro schemes, ten major development projects were largely accomplished. The Highlands acquired over fifty dams of varying sorts, almost as many power stations, and many miles of pipeline, tunnel and aqueduct. Some old lochs had been enlarged, drowning many acres of land, and some new ones had been created; and the glens and moors had been strung with a network of power lines under files of pylons, marching in spaced columns. (See Appendix.)

All the hydro schemes consisted of the main elements of dam, aqueduct, pipeline, tunnel, power station and transmission system in varying combinations. In each one, the most obvious and impressive features, as far as the public was concerned, were the dams. The largest dam is the one on the east end of Loch Mullardoch. With a length of 2,385 feet and a height of 160 feet, it needed 286,000 cubic yards of concrete to complete the two long wings of wall meeting on a small island at the outlet from the loch. Mullardoch is a mass gravity dam, depending on its bulk simply to sit in place, holding back the some seven and a half million cubic feet of water accumulated in the enlarged loch behind it. The second biggest dam, a twin structure on the Orrin River built in the third phase of the Conon Valley

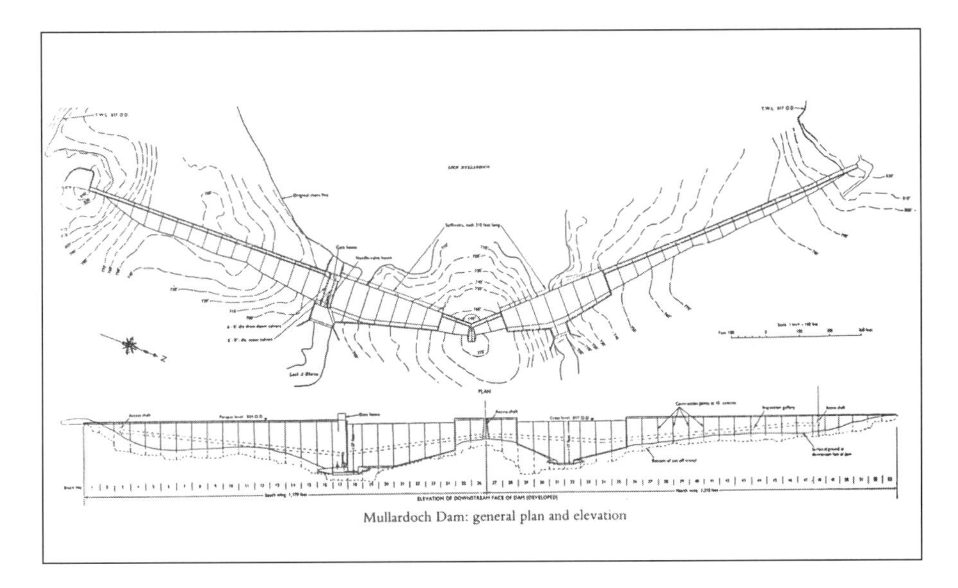

Mullardoch Dam: general plan and elevation

Fig. 1.
General plan and elevation of Mullardoch Dam, the largest built by the Hydro-Electric Board (*reproduced with permission from 'Special features of the Affric hydro-electric scheme (Scotland)', C.M. Roberts,* Proceedings of the Institution of Civil Engineers, *1953, Vol.* 2 (*1*)).

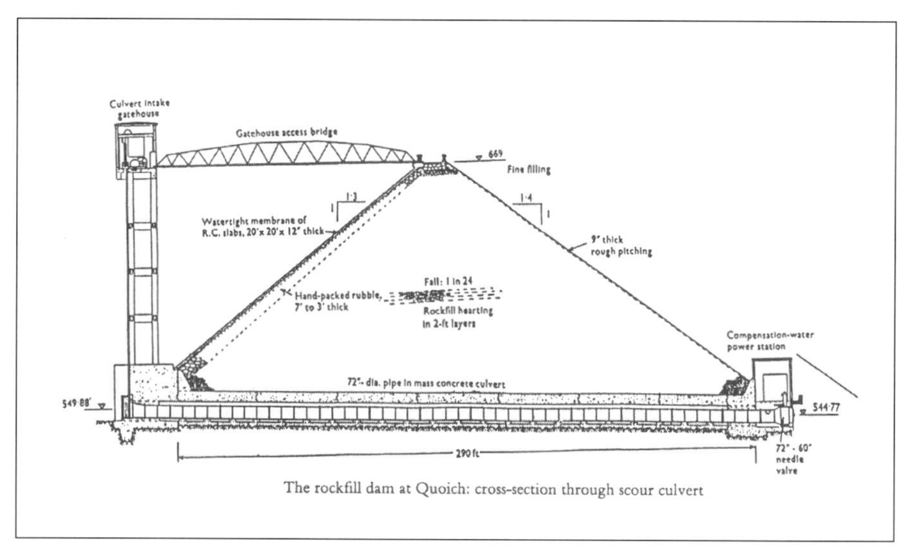

The rockfill dam at Quoich: cross-section through scour culvert

Fig. 2.
A cross-section through the Quoich rockfill dam (*reproduced with permission from 'The Garry and Moriston hydro-electric schemes',* Proceedings of the Institution of Civil Engineers, *1958, Vol. 11*).

scheme, is also a mass gravity dam; at 167 feet, it is slightly higher than Mullardoch but is only 1,025 feet long.

Anyone pausing at the Aultguish Inn on the long, lonely road across the heart of Ross-shire between Garve and Loch Broom will find the western horizon ruled off by the concrete wall of the Glascarnoch dam built across the Glascarnoch River. It required 186,000 cubic yards of concrete to make the 1,753-foot length of this view-blocker, that sustains the artificial 4.5-mile-long loch behind it, but much of its structure was made with earth fill. Its companion dam, on the Strathvaich River, another tributary of the Blackwater a few miles to the north-east, is also an earth-fill dam, with a concrete core, a type of construction clearly seen from its gently sloping sides. The Quoich dam is another rockfill giant, 1,050 feet long and 125 feet high; it penned back the waters of Loch Quoich and raised the surface by 100 feet, increasing the area of the loch from three to seven square miles, and necessitating the building of two more dams at the west end to stop the water spilling in that direction.

The most elegant of the major dams is also one of the most remote. Tucked away at the head of Glen Strathfarrar, twelve miles west of Struy Bridge, the Monar double-arch dam, the first of its type in Britain, is a relatively thin, concrete wall, gently curved from base to spillway and more spectacularly bowed from end to end. The narrow, steep-sided shape of the gorge where the River Farrar flowed east from Loch Monar allowed this design which needed only 36,000 cubic yards of concrete and cost nine per cent less than a mass gravity design doing the same job.

The Labour government of Clement Attlee, elected in 1945, made nationalisation of the country's larger industries a major part of its policy. The Central Electricity Board had been formed in 1933 on completion of the first

Plate 15.
The dam on Loch Benevean, Glen Affric (*author*).

Plate 16.
The downstream face of the Quoich dam, showing the slope of the rockfill structure (*author*).

Plate 17.
The double-arch Monar Dam at the head of Glen Strathfarrar (*author*).

national grid to operate all power stations, apart from those owned by the various town and city councils in Britain. The nationalisation of the electricity industry was placed under the direction of the Minister of Fuel and Power, Emmanuel Shinwell, MP for the Seaham district of Durham, a fiery left-winger who, in his earlier days, had been one of the Red Clydesiders. The North of Scotland Hydro-Electric Board feared that it would have to surrender its generating functions to the Central Electricity Board, an arrangement that would effectively reduce the Hydro-Electric Board to merely a distributing authority and bring an end to the ideal of using the Board as an instrument of economic regeneration in the Highlands. During 1945 and the early months of 1946, the two Boards argued over details of pricing and supply: some of the disputes related to technical matters but, on the issue of price, the Hydro-Electric Board adamantly resisted supplying the CEB at a rate that would reduce their income and endanger the development of uneconomic hydro-electric schemes in remote parts of the north.[63]

Emmanuel Shinwell came up to Pitlochry to meet Tom Johnston on the site of the Tummel-Garry scheme. At the same time, Herbert Morrison, Lord President of the Council, and the man in charge of the overall nationalisation policy, visited Loch Sloy. It was a perilous moment for the Board. After the traumatic Tummel-Garry inquiry, it was once again under threat, and this time the threat could result in its emasculation. Shinwell, Johnston and their respective advisers talked through the matters at stake. Hamish Mackinven takes pleasure in telling what happened next: 'The two old boys [Shinwell and Johnston] said to their officials they were going for a wee walk. They were both ex-Red Clydesiders, they had known each other for many years, they were both steeped in political guile. They wandered away down by where Loch Faskally would one day form. They were dressed the same way – black coats with velour collars, homburg hats. None of the nail-biting officials waiting and watching could hear a word that was being said. They came back and Shinwell said "I've decided the Board will retain its autonomy". That was it – and I often think that the birch trees by the loch when they are whispering in the wind are recalling that private conversation.'[64]

In May 1946 Shinwell told the Cabinet his decision and, in the following January, the Bill for the nationalisation of the electricity generating industry established that the North of Scotland Hydro-Electric Board would be given responsibility for the whole of the north of Scotland, an area of 21,638 square miles with a population of 1,165,608 people and including the cities of Dundee and Aberdeen.[65] The Grampian Electricity Supply Company and all the smaller private and municipal generating operations would be swallowed by the Board, whose borrowing facilities were to be increased from £30 million to £100 million.

The debate over nationalisation was given extra point by the severity of the winter in 1947. Crippling frosts settled over the Highlands and ice floes with seals aboard were observed in the Beauly Firth. Electricity supplies from the coal-burning stations were disrupted, and troops were deployed in February to move coal from the pitheads. When the temperature in Perth dropped to 12°F, parts of the Tay froze and workmen had to resort to picks to break frozen snow from the pavements, the *Perthshire Advertiser* carried a leader on the vital importance of the hydro-electric schemes.[66] Schools were closed, travel was disrupted, the harbour at Perth froze into immobility, and fifty sheep died when a south-bound train was trapped by a blizzard at Dalwhinnie.

In the midst of this austerity and finger-pinching cold, the prospect of cheap, abundant electricity from their own lochs and rivers sent a wave of anticipation across the Highlands and Islands. The excitement was encouraged of course by the Board, who mounted exhibitions about the new source of energy and whose switch-on ceremonies were already becoming a familiar event in village life.

On Tuesday 6 May 1947, Lady Mackenzie, the wife of Sir Hugh, threw the switch at Kyleakin that provided a small part of Skye with its first public electricity supply. 'Slowly but surely', cheered the *Inverness Courier*, 'electricity is being brought to remote glens and to the islands of the West, and that cannot fail to arrest depopulation and to make a positive contribution to the development of the Highlands.'[67] Sir Hugh said that his wife had asked him to appeal to 'the ladies of Skye' to make the best use of electricity. He also informed his audience that £2,000's worth of electric cookers had been sold in Mallaig. Five houses in Kyleakin received electricity at once and more than twenty others were wired and eagerly waiting to be connected. Poles were being erected along the coast to bring current to Broadford. An exhibition of appliances was mounted for three days and a Miss Scott, described as the Board's cookery demonstrator, showed the housewifes of Kyleakin how to use an electric cooker to grill steak and liver, and bake cheese scones, date slice and drop scones. The submarine cable from Kyle of Lochalsh was designed to bring power to 85 per cent of the Skye people but Edward MacColl happily announced after the opening ceremony that the Electricity Commissioners in London had just approved the Storr Lochs generating scheme. The details were published a month later. A dam was to be built at the north end of Loch Leathan, whence a pipe would convey the water to a power station near the mouth of the Bearreraig

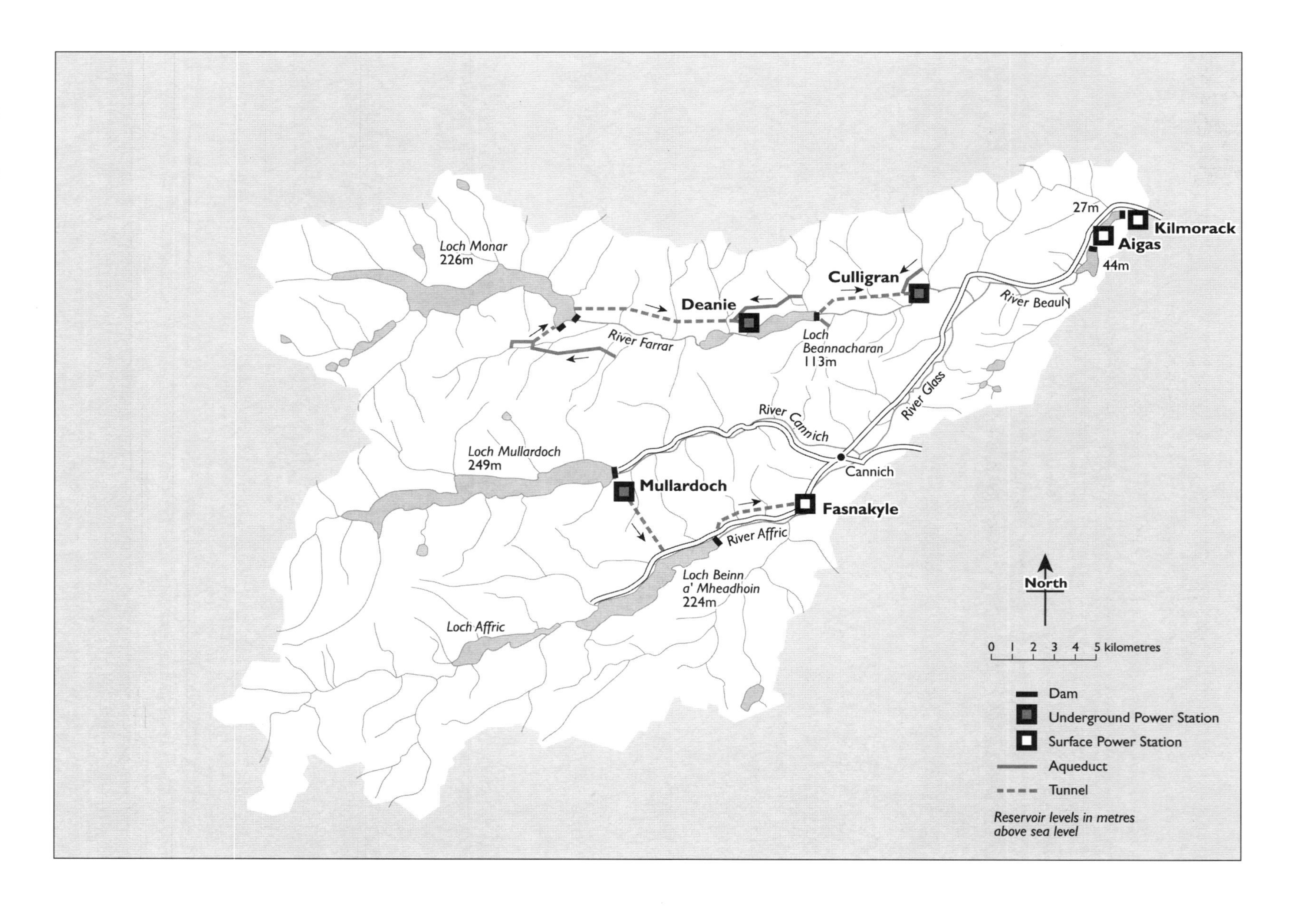

Loch Monar
226m
Deanie
Culligran
27m
Kilmorack
Aigas
44m
River Beauly
River Farrar
Loch Beannacharan 113m
River Glass
River Cannich
Cannich
Loch Mullardoch
249m
Mullardoch
Fasnakyle
River Affric
Loch Beinn a' Mheadhoin 224m
Loch Affric
North
0 1 2 3 4 5 kilometres
Dam
Underground Power Station
Surface Power Station
Aqueduct
Tunnel
Reservoir levels in metres above sea level

River. The project would cost £247,000 and would provide 5.5 million units per year to supply the island's 10,500 inhabitants.

At the end of May, Mrs Johnston was the central figure in another ceremony – the inauguration of the Glen Affric scheme. It was a glorious day, the sun shone and the trees were splendidly green. Close to four hundred people came to watch the proceedings and were entertained before and after the high point by the pipes and drums of the Queen's Own Cameron Highlanders. The Revd Angus Macleod opened the ceremony with a prayer and, after speeches by Edward MacColl, Tom Johnston and Cameron of Lochiel, Mrs Johnston pressed the switch on the Union Jack-draped table to fire a symbolic blast of explosive. A puff of rubble rose into the air from the banks of the river. [68] Along with the other pupils at Struy school, Archie Chisholm, then about thirteen years old, was taken on an outing to see the ceremony and recalls the pile of quarry dust and the puff of smoke when the gelignite charge was fired. Luncheon was then served for around two hundred guests in the canteen and recreation hall of the newly erected workcamp, big enough overall to accommodate two thousand men and put up in a week. The menu featured Strathglass salmon, and Lord Lovat and Sir Murdoch Macdonald MP delivered speeches.

Map 3.
Affric/Beauly

It was a moment for relaxation and celebration. The Board had survived the difficulties of its early years and could now get on with its aims. The shot heard in Strathglass that Friday could fairly be claimed to be the starting gun for the first major experiment in Highland development for many a long year.

[2] '... a cold job it was ...'

Some of the large hydro schemes employed up to two thousand workers, although there were rarely as many people on any site at one time. Most of them were men but a relatively small number of women also found work in the camps and various offices. Among the thousands could be found experts in almost every trade – joiners, engineers, fitters, mechanics, crane drivers, explosives handlers, drillers, cooks, office clerks, electricians, surveyors, all supported by a host of semi-skilled labourers who plied the pick and shovel wherever muscle power was needed. Some had no trade at all, save their wits, and offered themselves for work in the hope they could pick up something to get by or escape the eagle eye of the foreman long enough to enjoy their share of the high wages on offer. Some were prisoners of war put to work while they longed for repatriation or displaced persons (DPs) carving out a new life in a new country. And there were the fly guys, the wide boys, who worked their scams.

As time went on, the proportion of native Highlanders among the labour force increased. The Board came in for some criticism over recruitment (although recruitment was usually the responsibility of the various contractors) and was accused of not using locals but this charge was not long justified. Of the 1,739 employed on the Glen Affric scheme towards the end of 1949, 80 per cent were Scots and half of these were from the Highlands; the remainder comprised foreign workers (ten per cent), Irish (six per cent) and English (four per cent), said Sir Hugh Mackenzie in a talk to journalists in December that year.[1]

In November 1953, G. D. Banks, a Hydro Board information officer, reassured a meeting of the Gaelic Society of Perth that the Highlander was playing his full part in the great enterprise. 'There is scarcely a Scottish clan that is not represented [in the labour force]' he said. 'You will find Highlanders supervising the construction of new dams and power stations, some of them young engineers who, but for the Hydro-Electric Board, would have been abroad developing the resources of some foreign country instead of their homeland.'[2] In 1954, the local employment offices in the Highlands were able to claim that half the number of men for whom they had found work had been placed in civil engineering jobs, mostly those linked directly or indirectly to the hydro schemes.[3] Most of the 250 men employed in drilling the Invermoriston tunnel in 1957 were Scots, if not Highland, with a few Poles, English and one Newfoundlander – Harry Mugford, described as a 'machine doctor', who had married an Inverness woman and settled in the town.[4] On the Strathfarrar scheme in late 1959, 85 per cent of the work force was local and most of the rest came from other parts of Scotland.[5] The employment prospects with the consequent boost to the Highland economy were very welcome. The Revd Roderick Fraser in Dingwall recorded in 1952 that '[The schemes] have brought employment to local men, semi-skilled and unskilled, and talk of heavy pay packets for overtime and weekend work is common but difficult to confirm'.[6] The unemployment rate in Inverness-shire fell to 1.9 per cent, representing 708 individuals, in the summer of 1954.[7]

In the early days of the schemes, however, there was a shortage of skilled labour and a high turnover. At Sloy, in October 1945, men were in such short supply that some two hundred German prisoners of war and displaced Europeans of several nationalities were drafted into service. The number of Germans rose to almost four hundred, outnumbering British workmen by around nine to one. The POWs were brought up each day by train from their camp at Garelochhead to disembark at the little station at Inveruglas and, while they laboured, for example, on the access road to the dam site, their two guards sometimes resorted to a handy pub to pass the hours until the return journey.[8]

Balfour Beatty and Co, the main contractor for the Sloy dam, had to put up with inexperienced workers who refused to stay on the job for long. In the first two years, there was a turnover of nearly two thousand men just to maintain an average of two hundred on site and, in this context of flux, the prisoners of war proved to be the most stable element. Edmund Nuttall Sons and Co. had the contract for tunnelling. Labour shortages caused them to abandon the usual twelve-hour shift in favour of eight-hour shifts, three a day, and in its wake this brought additional accommodation problems. To overcome the lack of experience in tunnelling of many of the workers, a 'school' was run to provide three weeks of tuition. The lack of skilled tunnellers also proved a problem at the start of the Glen Affric scheme in 1949; volunteers from Lithuania were drafted and proved to be apt learners.[9] At Glen Affric 'there had been a turnover of 9,000 men ... the bulk of [them] had to be trained and sometimes they lost 30 per cent in a month', reported a spokesman for the contractor, John Cochrane and Sons, in December 1949.[10]

Although they may have constituted only a fraction of the labour force, the Irish achieved a prominence belying their actual numbers. The author of the description of the parish of Contin, where the Conon schemes were being constructed, recorded in the Ross and Cromarty volume of the *Third Statistical Account*: 'In the 1951 Register of Electors for Fannich and Grudie Bridge such names as Doohan, Gallagher, McCafferty, McCallion, McDevitt, O'Brien, O'Donnel, O'Rourke and Sweeney reveal the numerous additions to local labour.' Perhaps it was because their presence led to such novelties as Beauly having for a while the furthest north pub to serve draught Guinness that they are remembered so vividly; or more likely it is because they were the descendants spiritual and actual of the itinerant navvies who had built canals, tunnels, railway lines and roads during the industrial expansion of the preceding two centuries, and came with a reputation.

They found their way to the Highlands from England or from the western counties of Ireland where economic conditions and poor prospects had made emigration to find work a tradition. 'It was the boys who went to Scotland who kept many a family on its feet, always sending a few pound to pay for something,' said Patrick McBride. The men took the night boats from Londonderry or Belfast to the Broomielaw, paying the single fare of 7s 6d (in the late 1940s). Pat Kennedy came from a small farm in the Rosses in the west of Donegal and worked as a shuttering joiner at Rannoch, Inveraray and Shira. 'The girls working in the camp at Butterbridge were mostly from Donegal,' said Paddy Boyle.

'There was nine in the family,' said Patrick McGinley, from Creeslough in Donegal. 'Life was very hard at the time [the 1940s]. It was a happy enough childhood but we had nothing. Everybody was the same, nobody had anything. Our father and mother worked hard, our mother especially, although I don't remember much about her as I was six or seven when she died. One of my sisters was only two at the time. I was fourteen when I left the school. It was bare feet in the summertime, and we got shoes in the winter.'

Patrick left school to cut turf and work for a farmer and, by the time he was sixteen, he had saved enough from his wage of 23s 6d a week to buy a bicycle. This allowed him to join a gang of men who went by bus (the essential bicycles went ahead in a lorry) to Kildare to work at the turf cutting.

'We had potatoes for breakfast, dinner and tea – that was all we got for the three meals. I was too young to cut turf but I got a job mending barrows, putting handles on shovels and spades, making tea, all that carry-on. I was the youngest of the crowd that went away. The rest were all grown men who cut the turf. I got a job anyway.'

Patrick's brother, who was a foreman fitter with the English contractor, Cementation, rescued him from the Kildare turf bogs to work on the construction of the Ballyshannon hydro-electric scheme. 'He put me on driving an electric derrick. I got £6 a week and thought I was in heaven.'

From Donegal Patrick travelled to England, moving around, learning fitting and acquiring skill as a crane driver. All the time news of opportunities flew back and forth along the grapevine and it was through some mates that he learned there was plenty of work to be had with the Hydro. 'We were young at the time, so we would head up to Scotland to see what it was like. In Edinburgh, the man in the A and M Carmichael office told us, "Aye, there's plenty work here. Head away for Inveraray. We'll give you a voucher to go on the train. Somebody'll meet you and take you out to the works".'

Dungloe is a few miles north of Cresslough and it was there in his father's hotel that the nineteen-year-old Patrick Campbell was pulling pints and washing glasses when, one Christmas, eight men dressed 'like they had stepped out of a Hollywood movie' came in and bought whiskey all round. There was a long-established tradition of emigration in the family: Patrick's great-grandfather had worked in America, and his father had done the same. Now the sight of young men, some of whom had been his school chums, returning from the Highlands with the heroic status of tunnel tigers and obviously with cash to burn stirred him to quit the ailing hotel and try his own luck. Borrowing money for the fare, he boarded the night boat from Derry to Glasgow, and then caught a train to Pitlochry, where he transferred to a local bus for the last leg of the journey to the camp at Dalcroy near Loch Tummel.

Paddy Boyle was one of a family of nine children in Gweedore, on the exposed north-western corner of Donegal, and when he reached the age of seventeen he too caught the Derry boat; he made his way to Dalcroy and was taken on by A and M Carmichael: 'A Dunloe man put me in the tunnel. He gave me a machine for boring but I didn't even know how to work it.' After childhood on the small family farm near Creeslough, Patrick McBride followed his father and brothers to work on construction jobs in Britain, on the building of a reservoir at Dalry and in Walker's shipyard in Newcastle-on-Tyne before moving north to the Clunie dam project. As a boy he had heard all about the techniques and problems of practical civil engineering from his father, who used to show his sons with matchsticks and a cake of butter the method of timbering a trench.

Hugh McCorriston, the eldest of a family of ten in Coleraine, County Londonderry, had a similar story. After going to England when he was eighteen to find work and to escape from sectarian troubles, he worked in the gas industry and with the Calendar Cable Company before having to return to Ulster when his mother fell ill.

'I came back to Ireland and plodded about, working on farms and that, for a while,' he said. 'My father was then transferred up to Loch Sloy. He was an ex-army man, a powder monkey, he worked with explosives. He got a job with the Hydro at Loch Sloy, in the tunnels there, and when that was finished he came home. He was home about ten days when he got a phone call from the Glen Affric project, that they wanted him over. He went as a tunnel boss and he took me with him, and that was it.

'We had a mixture of Irish, Scots and Polish – one or two other nationalities but it was mostly that three. The Scots were not at all pleased to be working in the tunnels – they didn't like it. Now most of them were coalminers and they were used to working in little, narrow seams, and when they came up and saw this [the tunnel] they said "Bloody hell", would work about a week and be off again. They couldn't cope with the space. It was funny, you'd think it would be the other way around but they were so used to the cramped conditions, and this was too big. I got on very well with the Poles. No problem. Quite a nice bunch of lads. Good workers. Quite a lot of them stayed on, got married and so forth. There are a few around Inverness yet.'

Otton Stainke came from a small village near Poznan where his family had a farm. His service in the Polish Army brought him to Scotland and in 1948 he began to work on the Affric scheme. Wodek Majewski had a similar background in the village of Czernno but found his way to the Highlands by a different route: 'I left home in 1943 and was working in Germany during the War. When it finished I was with the Americans for about a year and a half, and came to Britain in 1948. We landed in Cambridge and then I went to Yorkshire [to the coalfields]. Then an agent came from Cannich, from Cochrane, and said he was looking for fifty men – strong men, and it's good pay and heavy work but, if you go, it's too far to send you back. We came up on the night train and landed in Inverness at six o'clock in the morning, and buses took us to Cannich. It was June, I think, and the weather was quite good. That was my first time in Scotland.'

'On the Quoich there were fair numbers of Poles and east coast men – from Buckie and that area – all great workers,' said Laurie Donald. 'There was a vast amount of overtime. They didn't get rained off, they would work away in the pouring rain. At Foyers the force was mostly Scottish. A fair number had become practised in tunnel work – the attraction was good money, good bonuses, long hours – but it was very hard work. At Quoich we had one German who lived in a caravan – a kind of a loner, in fact, he worked alone maintaining pumps, doing his own thing all the time. We had an ex-German POW at the Orrin – he worked in estimating – and we also had the Army. In those days they ran a scheme whereby engineering officers would be put out to work with contractors for free for the experience.'

Paddy Paterson remembers the DPs: 'They all fitted in – Latvians, Lithuanians, Ukrainians. There were some Germans as well. There was a chap called Heinz, an ex-POW who worked in the black gang [the fitters and mechanics]. Many of them became naturalised Britons when the schemes were done.'

'There was even an Indian at Cruachan,' said Barry McDermot. 'Big Singh – he drove one of the big dump trucks. The spoil from the tunnels was used to widen the road along Loch Awe. One night Singh was dumping and he went too near [the edge] and when he tipped up the whole lot went over, but Singh was fast enough and got out, but the truck was lost for ever, because it is a wild drop there to the bottom.'

Among the Scots who joined the schemes was Donald Macleod. Demobbed in 1948 from his National Service with the RAF, Donald returned to his native island of Harris and found work building houses. That came to an end in 1950. 'I never had unemployment benefit in my life', he said, 'and three of us decided to head off down to Pitlochry. Fortunately at that period there was loads of work, you could find work anywhere, any day. So, that's where we went and we got started the next day.'

Before the War, Bill Mackenzie, the son of a crofter at Fairburn at the east end of Strathconon, cycled every day to do construction work for the Air Ministry at the Invergordon naval base from his home in Muir of Ord. In 1940 he enlisted in the Royal Engineers, volunteered for bomb disposal work and, when the War ended, returned to Muir of Ord. He worked for a plumbing firm for five years and then joined Duncan Logan Ltd on the schemes.

'If the hydro board hadn't got going when it did we might have had a replication of what happened after the First World War,' said Iain Macmaster, a native of Knoydart, who learned electrical engineering with James Scott and Company before working on the schemes. 'All the men came home from the forces and in the Highlands there was nothing for many of them to do. Many more would have opted to leave but the money they earned on the schemes gave them a chance they wouldn't otherwise have had. It also gave them technical skills and experience of industrial work.'

William Rosie, from John o'Groats, said, 'After the War work was kind of scarce up here, so I went down to the hydro schemes and I was there for over two year in Cannich. There were quite a few men from Caithness.'

Don Smith was another who came to the schemes after a period in the armed forces. He had left school in his native Newark-on-Trent at the age of fourteen and had begun to serve his time as a toolroom apprentice before enlisting in the Royal Navy in April 1942. The Navy turned him into a qualified electrician and, after demob in June 1946, he found work with Balfour Beatty and then with the Air Ministry as an electrical fitter. In the summer of 1948, and by then married, he learned there was 'good money on the dams in the Highlands – £12 a week. I came up for a week [to Inverness] and got a job'.

Iain MacRae came home from two years in the RAF in 1954 and re-joined the firm in Beauly where he had served his apprenticeship as a joiner before being called up. Shortly afterwards, the boss died and the firm was taken over by Duncan Logan Ltd, the construction company based in Muir of Ord and then expanding rapidly with contracts all over the country. One of these was for the building of the new dam at Invermoriston, and Iain MacRae was sent there as a joiner. In his army service, Don West drove trucks between Benghazi and Tobruk and it seemed natural enough that, after returning to his home at Struy Bridge in Strathglass, he should drive trucks for John Cochrane and Sons in the last days of the Glen Affric scheme and later find a similar post with Duncan Logan Ltd.

Sybil Davidson, who had learned secretarial skills at school in Elgin, began working for John Cochrane and Sons in 1947: 'I can't really remember how I ended up applying for the job in Cannich. I had an interview, and may have applied through word of mouth. The firm's office was in Church Street in Inverness, and I was petrified. I can't remember the interview but it must have been all right, I got the job.' Mairi Stewart, whose family had moved from Skye, began work in the same office at the same time: she had just finished training in the secretarial college in Inverness when her father met on the bus a man who suggested she try for a job at Cannich.

Students, too, found vacation jobs on the schemes. Antoin MacGabhann and five companions caught a night boat from Dublin to Glasgow in 1955 and landed work on the Shira scheme near Inveraray. 'It rained all day, every day, or so it seemed, in the mountains,' he wrote. 'We were not provided with any raingear, and slipped and slid as we helped others to push a large pipe up the mountain to a prepared trench. We did our best to impress on our first few days anyway. Perhaps we hoped to hide in the pipes later. But a big Donegal ganger told us to go down to the office "to get our cards". We didn't know what this meant but went down the mountain and got our cards. We climbed up again and innocently handed our cards to the ganger. He reacted angrily and said "What the hell are you doing back here? You're sacked" ... When we asked him why, he said there was no point in other people pushing the pipe up the mountain and us "Fucking looking at them".'

MacGabhann found another job on another scheme where conditions were better. Raingear was supplied and he became a scaffolder's mate, scaffolding the inside of a shaft. 'One day I emerged into daylight at the top and stepped on a rusty nail. I worked on but after a few hours I couldn't walk at all, and the ever-present ambulance carted me off to a doctor for a tetanus injection. I spent three days in bed, not seeing anybody between 7 a.m. and 8 p.m. The shifts were twelve hours and ... my friend used to leave bread and water with me before going off to work ... I enjoyed my stint on the tunnels and got on well with my immediate boss, scaffolder Victor from Northern Ireland. I saw no rowdiness or drunkenness, everybody just worked and slept, and would queue up every week to wire home money to their families. I saved about £60 and took the overnight bus from Edinburgh to London ... and blew the £60.'

Throughout the whole period of the construction of the schemes, the workforce was fluid. Men came and went. Gangers tried hard to keep good men in their squads but finally there was nothing a foreman could do to stop one of his workers picking up his cards and his gear and going down the road.

'We had a lot of Harris and Skye men,' recalled Bill Mackenzie, a ganger on the Meig and Orrin dams. 'The Harris men stuck it out and there are a lot living in the village here yet [Muir of Ord, Ross-shire]. There were some Irish as well. I remember a Paddy Doyle, a fine chap, a big tall man. He always carried his jacket with him even for short distances and I says to him one day "Paddy, why are you guarding your jacket?" "Here, Scotty, I'll show you," he says, and he had in it an old Warlock tobacco tin full of pound notes. Anyway, Paddy was with the squad down about seventy feet in the rock, in the cutting, on the Meig. We went up and down ladders. There were no guard doors on the top and when it started to rain the pebbles were coming down. No hard hats. You would usually hear the stones coming and you would stand against the rock. This day Paddy never heard them and one nailed him and drew blood, gave him a good knock. He never said a word but went over and lifted his jacket: "Cheerio, Scotty, I'll be seeing you", and he just left.'

The surge of construction in the Highlands influenced youngsters and showed them new possibilities, new career opportunities. As a schoolboy from Gairloch, boarding on the east coast in Dingwall, as was then the norm for many from the West Highlands during their secondary education, Roy Macintyre saw the work at Grudie Bridge, Glascarnoch and in his native district: 'All this activity in our own area interested me as a boy. It was something completely new and it sowed the seed in my mind of becoming a civil engineer. One of the engineers on the Kerry Falls scheme at Gairloch was a young fellow called Roy Osborne. He became friendly with my sister and subsequently they married. I remember him taking me up to the Falls and I was thinking they were great guys who had the chance to be involved in this kind of work. I went to Glasgow University and did civil engineering. At that time it was a sandwich course and we used to find work in the summer, from May to September. In my first year, the summer of 1954, I got a job with Duncan Logan Ltd and of course being a first-year student I guess I was a bit of a liability, with no practical experience. I remember I was given £1 a week and I think they paid my digs in Dingwall, and I got a job as a student engineer on the Meig Dam. At first it was a steep learning curve but as the weeks passed I became more useful, I learned to do setting out – in reality I think I learned more in a few weeks there than I learned in the first year at university. I loved it and got to know the men and the activity. The whole thing was fascinating to me.

'I was a pretty green engineer at first but you didn't take long to pick things up. The engineer on the Meig was an interesting fellow called Tom Critchley, an Englishman who had come up to Gairloch, giving up his career to start a crofting life. He was probably a bit of an idealist and I don't think he would have stuck it long but fortunately Willie Logan arrived in Gairloch almost in the same year and needed an engineer, a site agent, to look after the works. Tom started a career with Logan that was to take him to many different dams. When I was on the Meig, Tom was the agent there; he had an assistant Jack Forsyth who had been at the Kerry Falls, and Tom and Jack were very helpful to a young fellow. I was given a lot of help and a lot of rope as well. Maybe they were at the age when they were glad to have a young fellow take on the graft of setting the thing out as it involved humping up and down the hills quite a lot.

'Life on the dam was great. We had our own canteen there, and there were a couple of ladies who served the meals. I stayed in Dingwall and I used to travel up every day. I got a lift from a Czech who was the site agent at the Luichart power station. His name was Wilson, which must have been anglicised, and he was a character, close on seventy years old at the time. Then I would walk from Luichart power station up the other side of the river where they blasted a road to the main dam in the Strathcarron valley. The camp was down at the power station.

'The workers pulled my leg a bit at first, but they were a great bunch and I think I was a part of the team. Catching the bus with a crowd of navvies on a cold Monday morning at six o'clock and there was just one adjective, the f-word, and you heard it so often you became immune to it and it had no meaning. It didn't have any meaning for them, each sentence would be liberally sprinkled with three or four uses of it.'

Before any building could start, before the tunnels, dams and associated structures could be designed, the country had to be surveyed in great detail. Roy Osborne helped to make a preliminary survey of the upper parts of Glen Strathfarrar between 1949 and 1951, as his first job after graduating in engineering from Woolwich College, London. His interest in hydro-electric engineering had been sparked when his father had bought him a book on the subject and, on graduation, he found employment with the consulting engineering firm of Sir William Halcrow and Partners. 'In those days,' he said, 'there was no knowledge at all about the Highlands in the south. I went to the library and looked at an atlas and there, on the last couple of pages, was Inverness-shire. I came up on the overnight train and, when dawn came, I was in the Cairngorms, seeing for the first time rivers with boulders. All the rivers at home had mud. I was interviewed by an old, very experienced engineer, Duncan Kennedy, in Halcrow's Inverness office. When I went to start work in Glen Strathfarrar, one of the engineers hired a car and took me up to the digs in a cottage belonging to a war widow at Culligran. Her husband's relations lived all around and we ate venison and salmon, we lived very well. Meanwhile the work was going on down Strathglass at the Mullardoch dam and there was another lodger, a fitter on that scheme.

'We got pretty fit, tramping about and wading the rivers. There were two of us [engineers] in the team, with two chainmen, and we worked from May until about November. It took some time as we had to survey for three dams. There was a sense of wonder. My shoes were no good on the hillsides, and I had to get boots with tackets. It was great fun, and it was all new. I remember the flies in the bracken, I don't recall the midgies. The aim was to make a contour survey of the [Loch Monar area] using Abney levels, which are held up to the eye. One chap I worked with used to tie a handkerchief around his head, go a certain distance, and then we would sight the handkerchief, and go beyond him and sight that. It produced a rough map. The Abney level worked like a theodolite and used a bubble to give a horizontal level. We started from the Ordnance Survey base levels on their six-inch maps. More exciting was the survey of the proposed tunnel line and the outfall for the pipeline to the power station. A detailed contour map was produced where it was felt to be necessary. The preliminary survey was

Plate 18.
Surveyors working on the site of the Glascarnoch tunnel portal, November 1951 (*NOSHEB*).

probably taken only as far as to work out the area to be impounded to make the reservoir. From this rough survey you could get a possible place for the dam. Then, precise surveys would be made of the rock, the depth of soil, and so on.'

On a year out between leaving school and starting university, Sandy Payne worked for six months as a chainman on the Glenstrathfarrar scheme. 'We kept regular hours. We left the camp at eight in the morning and went up to the dam in Land Rovers; if the weather was poor you might come back a bit early, but sometimes we worked late and got overtime. It was a pretty good wage, especially for someone just out of school. As a chainman I was climbing around the dam as it was being built, holding the staff, making marks on the cement so the engineers could take levels. There was a group of four or five chainmen and we had our own clique and our own way of life. There were a few perks – we had a fresh supply of bread, sugar, milk and biscuits – only for the chainmen. I don't know how it came about but we were very jealous of our perks and nobody would intrude on them. We had our own room, just a cupboard in one of the huts, with a gas stove and a kettle, self-contained, a place where when we weren't needed, we could sit around drinking tea, eating big slabs of cheese (another perk) and playing cards. We felt quite privileged although we must have been the lowest in the pecking order.

'When the engineers found that I intended to go to university to study maths, they got me into doing their calculations for them. They were having difficulty with the equations for the double-arch structure [of the Monar dam] and I was given the job of working through and checking their figures – slide rule and log tables in those days. I was peeved at first because I wanted to be the hard engineer, out climbing over the dam, but when January came I was quite pleased to be able to sit inside doing calculations. Can you imagine someone like me being given paper, a pencil and a rubber, and a series of equations, and being asked to work out where the next layer of cement should go? I felt the responsibility but eventually I got the confidence to carry on with it. It was all checked. I feel quite privileged to have been part of the building of that dam in more ways than one'.

Sandy Payne spent most of his spare time walking around the beautiful hills, sometimes climbing in rubber boots and overalls, and observing the bird life: 'It was an eye-opener being stuck in a glen like that. Fantastic, mind-blowing – with golden eagles, crossbills, greenshank.'

As the chief engineer with the Mitchell Construction Company on the Killin-Lochay tunnel system in Breadalbane, Bob Sim took a leading role in surveying. The aim was to prepare a detailed map of the rugged hill country which would then be used to ensure the tunnellers stayed precisely on course. Tunnels were usually driven from both ends, and sometimes from additional internal faces accessed by adits, or side tunnels; and the expense if two sections happened to miss each other by passing to either side or above or below would have been almost as great as the embarrassment.

The survey team, often four engineers and two chainmen, began by establishing a baseline 3,000 feet long. This was laid out on the level floor of a strath according to theodolite bearings and measured manually with a tape made from Invar steel, according to theodolite bearings. Each 100-foot section was marked by driving in an oak post and cementing it in place in the gravelly moraine. Nothing was spared in the search for accuracy: the steel in the tape was specially selected for its low coefficient of expansion and the temperature was always recorded during its use; at each end of the 100-foot section the tension in the tape was adjusted for the catenary, or sag, by suspending weights over pulleys, and sometimes it was necessary to dig a

River Lyon
Loch an Daimh 433m
Loch Lyon 343m
Lubreoch
Cashlie
Stronuich Reservoir 292m
Lochan na Lairige 521m
Kenmore
Loch Tay 106m
River Lochay
Lochay
Finlarig
Killin
River Dochart
Loch Breaclaich 443m
Lednock
Loch Lednock 352m
St. Fillans
Lochearnhead
Loch Earn 97m
Dalchonzie
North
0 1 2 3 4 5 kilometres
Dam
Underground Power Station
Surface Power Station
Aqueduct
Tunnel
Reservoir levels in metres above sea level

trench to allow the tape to sag freely; further corrections were made for the slope of the ground, the four-inch width of the oak post and the curvature of the Earth. Then the measurement of each section was repeated from the other end, and this was done a further three times until for each 100-foot length of baseline sixteen observations had been completed. A range of 0.02 inches in the results was considered acceptable.

Map 4.
Breadalbane

From the baseline, a series of quadrilaterals measured by microptic theodolites were drawn across the hills from one ridge to the next. At the end of the series a final check baseline had to match exactly the length calculated from the initial baseline. From the corners of the quadrilaterals very accurate theodolite bearings were taken to pinpoint the positions of the tunnel mouths; these bearings had to agree within three seconds of arc per 180 degrees. Altitudes were monitored by measuring levels down one glen and up the next.

Weather could foil the surveying as it could affect much else. On one scheme on Ben Vorlich, Bob Sim climbed up five times before he managed to obtain a mist-free day and eventually achieved success only by bivouacking on the summit in a sleeping bag.

'The helicopter revolutionised surveying,' explained Bob Sim. 'Although it was expensive to hire and its use had to be carefully planned, it saved a great deal of time. We were the first to use a helicopter – a Bell, with a glass bubble, open sides and a skeleton tail, the same machine as in the television series *Whirlybirds*. We hired it from a Derby firm for ten days and used it to fly the concrete columns, with instrument mountings for our surveying, up to the observation points we had established on the ridges. We made seventy-five flights in ten days and erected twenty-seven observation points. The helicopter could lift only 400 pounds weight and the workmen still had to walk up to the sites but they were spared the need to carry so much material.'

Another advance at around the same time was to use a computer to calculate the quadrilaterals covering the course of the tunnel. The standard method using eight-figure logarithmic tables could take ten days of office work. 'The computer was in Derby, in an aircraft hangar, and it looked like a series of huge grey wardrobes,' recalled Bob. 'It was owned by English Electric and it did the calculations for us by return of post.'

The detailed design of a dam begins with knowledge of the foundation on which the structure will rest. 'In the places we worked, you can imagine the rock was pretty near the surface, so the procedure was to take off all the peat and overburden until you reached solid rock,' said Roy Macintyre. 'The rock was cleaned, so clean you could eat your dinner off it, it was really pressure-hosed and scoured, and then there would be an examination of the rock. In the case of the Monar dam there was an extensive programme of grouting because the geologists found a faultline and it was thought to be potentially quite serious. They spent a lot of time doing what they called stage grouting. I think there were three stages. The grouting was fibrillated into the rock maybe for 50 or 100 feet, pressure grouting, and then they would drill again and do a second stage with more liquid grout the second time, and they did it again – by doing it three times they reckoned to seal all the fissures. I think we did grouting on the Meig dam as well. On the Meig, from memory, it was done after the dam was built. It certainly was on Torr Achilty dam – by drilling right down through the dam and into the rock, and we did pressure grouting through the dam, as an extra security.'

A rockfill or earthfill dam needs quite an elaborate cut-off trench deep into the ground under it, pressure grouted and fitted with a 'blanket' to stop the water coming up. Rockfill dams are massive structures, with a 30° slope on both faces, and the upstream face has to be protected from wave action with concrete slabs. A concrete dam also has a cut-off trench to stop the passage of water under the dam, pressure grouted and possibly twenty feet deep to ensure the structure is resting on solid rock. If the rock is badly fissured, the pressure from water seeping under the structure could threaten the dam with uplift. 'A dam has the smallest factor of safety of any civil engineering structure, because it rests on stability,' explained Roy Osborne. 'It's in balance under its own weight. Only uplifts can put it wrong, or if you get a flood and the water rises too high over the dam, to create a greater overturning pressure. This defines the spillway length.

'The front-face slope of a dam is about 1 in 10, and the back face about 10 in 7. This directs pressure downwards to increase stability. The rule is to keep the resultant thrust, the resultant of the horizontal push of the water and the downward push of the weight of the dam itself, falling through the middle third of the base, or the 'footprint', of the dam. If the thrust goes outside that, the dam will become unstable and will topple. The pressure is dependent on the depth of the water. We had to calculate what we thought the largest flood from upstream might be, and the dam crest had to be large enough to take it. In the days before computers, it required a lot of tedious calculation.'

The leakage of water through a dam through the joints between the concrete blocks is prevented by a water stop, sections of copper, one eighth of an inch thick, anchored into the concrete so that, when the joints expand, the copper moves and still keeps the water out. This is a source of failure in many dams. Some stable dams can show slight leakage through the joints.

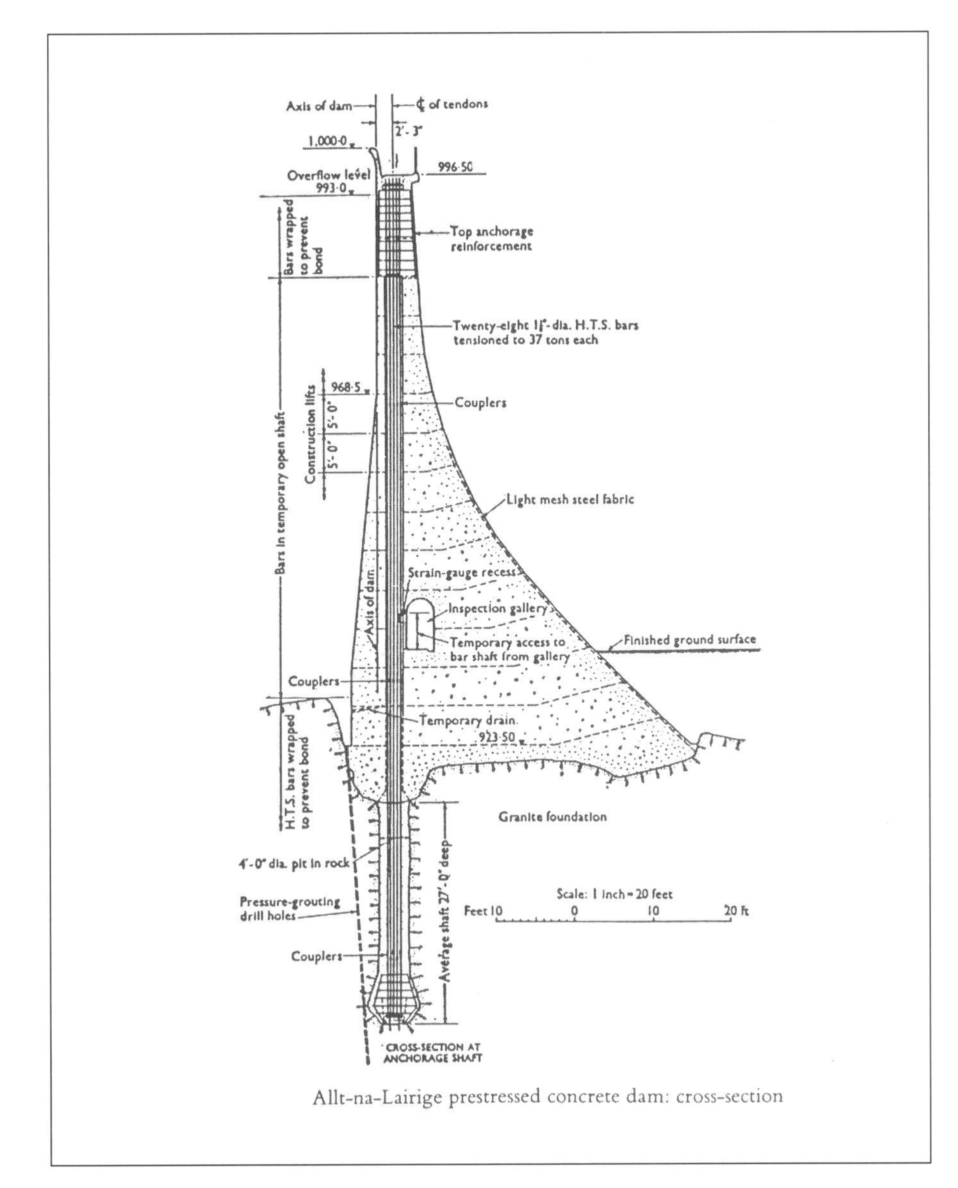

Allt-na-Lairige prestressed concrete dam: cross-section

Fig. 3.
A cross-section through the prestressed dam at Allt-na-Lairigie. This is the only dam of its type built by the Hydro-Electric Board (*reproduced with permission from 'Allt-na-Lairigie prestressed concrete dam', J.A. Banks, Proceedings of the Institution of Civil Engineers, 1957, Vol. 6*).

The contraction of setting concrete is the big movement. After that, there may be slight movement, enough to cause some leakage, from temperature changes but these changes are very small and dams are usually extensively grouted along the contraction joints to protect the interior from external temperature fluctuations.

As the schemes progressed, there was a degree of experimentation. 'Dam concrete is in compression and the dam is designed on that basis, narrow at the top and wide at the base, stable under its own weight,' explained Professor Ronald Birse. 'Thin-arch dams, where the upstream curve adds strength, can be built in a narrow valley. There's at least one pre-stressed dam in the Highlands. High tensile steel cables were put in and these were

Plate 19.
Clunie dam, November 1948. The north side of the dam under construction, as the river rushes past beyond the coffer dam (*NOSHEB*).

tensioned so that they hold the dam down to eye-bolts grouted in the rock and are literally tightened up at the top, stretched like an extremely strong elastic band. Dams are designed virtually to last for ever.' The pre-stressed dam stands on the Allt-na-Lairigie, a tributary of the River Fyne. The decision to convert what was first conceived as a conventional mass gravity dam was taken to allow building a dam fourteen feet higher than the original, an increase that permitted a greater storage capacity and a more dependable power output. There was an element of risk in the change of plan, as it would entail building the first dam of its kind in the world. This kind of challenge, however, was meat and drink to Angus Fulton, the Board's chief civil and hydraulic engineer, and work began on the Allt-na-Lairigie dam in 1953. One hundred and thirty-eight high-tension steel bars were placed inside the vertical dam wall and anchored to the granite foundation. The scheme opened in 1956.

After his initial surveying in Glen Strathfarrar and a period of National Service, Roy Osborne continued working for Halcrow on dam design. A new interferometric technique for the surveying of a line had become available and he tried it out on the then newly-completed Mullardoch dam. The purpose was to measure the deflection of the dam after the impoundment of the mass of water behind it. His measurements showed that the structure had moved uphill by a fraction; the valley floor had sunk and the dam had tilted slightly, only thousandths of an inch but still no one could quite believe it had happened.

Economics always came into the equations for dam design and construction – the final aim, after all, was to generate electricity at the most advantageous cost per unit. At the same time the unique combination of opportunities and challenges at each site had to be taken into account. The contractors on the schemes built up over the years an impressive expertise in dealing with recalcitrant rock, unexpected faultlines and new materials.

In the beginning, though, the work could be very tough. Patrick McBride was among the small group of men who launched the transfer of the Tummel-Garry scheme from drawing board to massive reality in the latter months of 1947. After they had completed enough of the work camp to allow the recruitment of more workers, by providing some place for them to eat and sleep, they turned their attention to the Tummel River, where it flowed from the east end of Loch Tummel through a gap in the Perthshire hills before being joined by the Garry and turning south past the village of Pitlochry. The men's task was to pile the fast-flowing river, the first stage in the construction of the Tummel Dam.

'And a cold job it was,' said Patrick. 'To pile the river and get all the

WIMPEY
WIMPEY
TUMMEL GARRY PROJECT
Contract 16 - Clunie Dam
View from Left Bank of area upstream
of L.H. Spillway Section
Ser. No. 18 Date: 5 11 48

Plate 20.

Left. Excavating a trench on the Pitlochry dam (*Perth Museum. Copyright: Louis Flood*).

Plate 21.

Right. Pouring concrete on the Pitlochry dam (*Perth Museum. Copyright: Louis Flood*).

water to one side. We erected two derricks. We piled the river and got the water diverted, and the next thing was the drilling and the blasting. My brother, he was at that now, blasting and all that business. By this time we had the camp well under control and a lot of people started coming, you know, these boys who travelled from one contract to another. Quite a few men came in a short period of time. We had some characters up there, I tell you.

'Then the Pitlochry project came on stream. After Christmas. The starting day for the Pitlochry dam was Easter Tuesday 1947. And you know that at Easter time we all go for a long weekend. But the boss, Mr Howard, came to me and said I had to be back for Tuesday morning for eight o'clock. I was the man he picked to go to Pitlochry to get six or eight blokes to be going on with. So I picked them out, and we arrived in Pitlochry and went in there at about a quarter to eight, and he was stood there with his watch. "So", he says, "that's starting time. That's the date the contract was signed for." The first job we had to do was shift a lot of footpaths that sightseers walked on,

Plate 22.
Engineers surveying at a tunnel entrance (*Perth Museum. Copyright: Louis Flood*).

put them away back [from the building site]. We built the office, the store, the staff quarters, and we got fed in two little canteens by other contractors. Anyway we got all that under control. The staff place was built, and the staff started to move in.

'The next job was the river. It was a big contract, piling that dam, putting up these derricks. We got things well under control but some time on, in June or July, a holiday period came on, the Scots trades fairs, and they all wanted holidays – but I worked there, stayed there 'til the rest came back. Then, a big flood came down from rain up north and did a quarter-million pounds' worth of damage in one night.' The flood on the Tummel in September 1947 breached the coffer dam, collapsed the left bank and toppled two cranes, but by 8 October the *Perthshire Advertiser* could report that all the damage had been repaired, in a week less than anticipated, and work had resumed.[11]

'The flood washed over the top of the piles, washed away the ground, washed down the derricks,' said Patrick McBride. 'All the timber we had sitting on the shore was washed downstream. That all spread out on the farmers' land and when the waters subsided there were big heavy timbers everywhere. Every day the farmers were coming up: "Get this timber the hell out of there." It cost more to pick it up than we could have bought it for, but it had to be done, we had to pick it up. Anyway, we got back on our feet again, got the dam sorted again, put the derricks up, started in the middle, drilling, blasting. Meantime my brother had come there and my uncle – he was works manager. So we carried on.

'What we called the North Cut ran out towards the village. [The North Cut marked the location of the dam's foundation on the left bank.] We found terrible bad ground and had to go a wild depth down. We sank piles for thirty or forty feet, but when we got down to the bottom we still hadn't got good ground. Then it had to be all heavy timber. There was a character there – we called him the Great Charlie Logue, he was an all-round heavy timber man – and he used to have a lot of long-distance followers who came with him. So the company got hold of Charlie and sent him up. Every day they were coming, a puckle of his mates, all timber men, and Charlie timbered that trench until he was down on the solid ground – down about fifty feet. All heavy timber along the sides – big boards, 12 inches by 12 inches, and 12 by 6. Then grout was pumped into it. That was the North Cut.

'Then we had the South Cut. Again it was very high ground. And there were a couple of very big buildings up on top, very nice property, that we weren't allowed to destroy. Again, Charlie Logue had the answer – dig it out from the bottom up. He and his crew put the tunnel headings in the bottom. When we had the first portion out, we put the boards in, they pumped in the concrete and when that was done, we took out the timber, and started again and came up another lift, and up another lift and so on, until we came to the top. Everything went very well after that.'

The pouring of the first layers of concrete on the dam began, working first on one side of the river bed from which the water flow had been diverted and retained behind a coffer dam. Temporary openings were built into the lower levels to accommodate the river when attention was switched to the other side of the coffer dam. When it came time to close the temporary openings, wooden gates were dropped into place and the gaps were filled with concrete.

Working in the rivers could present novel situations. In Glenmoriston a driller was assigned one day to drill a large rock obstructing the flow of the river to prepare it for blasting. A crane lowered him onto the rock but, as he drilled, the bit slipped and he fell into the freezing water. 'I was down at the workshop when I heard this clitter-clatter like a music box coming down the road,' recalled Iain MacRae. 'I came round the corner and here was the

Plate 23.
Brushing down and finishing off the concrete face of possibly the Pitlochry dam (*Perth Museum. Copyright: Louis Flood*).

Plate 24.
Glen Errochty, 1950. The foreman, Sandy Brown, beside one of the skips for carrying concrete from the batching plant to the dam (*Donald Macleod*).

driller. He had icicles hanging round him and every step he took he was going tinkle-tinkle. He was on his way down to the hut to get dried off and go back to work again.'

Up in the hills of the glen of Errochty, on the Trinafour dam in the Tummel-Garry scheme, Donald Macleod joined a team in charge of mixing and pouring batches of concrete. 'I was on the batching plant when they were building the actual cement mixer housing and the conveyor belt that fed the mixers,' he said. 'Besides having a better number there was also more money in it. At that time I was picking up £13-14 a week which was big money in those days. So, that was at the initial stages of the Errochty dam, they were tearing out the base of the river right across the valley, installing the blondin [the overhead cableway][12] to carry the concrete out to whatever point it was required. I was working with A and M Carmichael, and it was hard.

'We worked 100-odd hours in a week. We started in the morning at half past seven, and we were never rained off. It would be bad weather and very little doing, and the job would be blown off, the actual job itself, and we would go away down to the checking-out office and we were told, oh you'd better go back, wait. We just sat around but we were never sent home, unless there was a threat of snow, for we were way out in the glen.'

The housing for the batching plant had a frame of four-inch steel girders that had to be bolted together. 'We had no safety helmets, no safety belts, no gloves, nothing,' recalled Donald Macleod. 'On a cold frosty morning, your fingers were sticking to the iron. They gave us no clothes at all. You had to provide for yourself. No boots, no donkey jackets. Now that steel frame was sixty-odd feet high. I just climbed up. We had problems with some of the girders. The holes were not fitting, you know. And we would argue and big Sandy Brown, the foreman, would start up to show us what was to be done, but he would change his mind. He was no climber. Climbing was no problem to me at all.

'Then, big steel containers were built from sheet iron to hold the different types of aggregate. There were eight of them and underneath that we had eight conveyor belts going out to the concrete mixers. Two big mixers. This hopper would be on the top for the cement which was wormed from a shed close by.'

'The boys in the cement shed had good money but they deserved it,' said Don West about the Loch Luichart dam. 'The cement came in bags, and they had to open them and tip maybe fifty bags at a time into the mixing. The dust and the sweat – you'd think their heads were harled – and they wouldn't

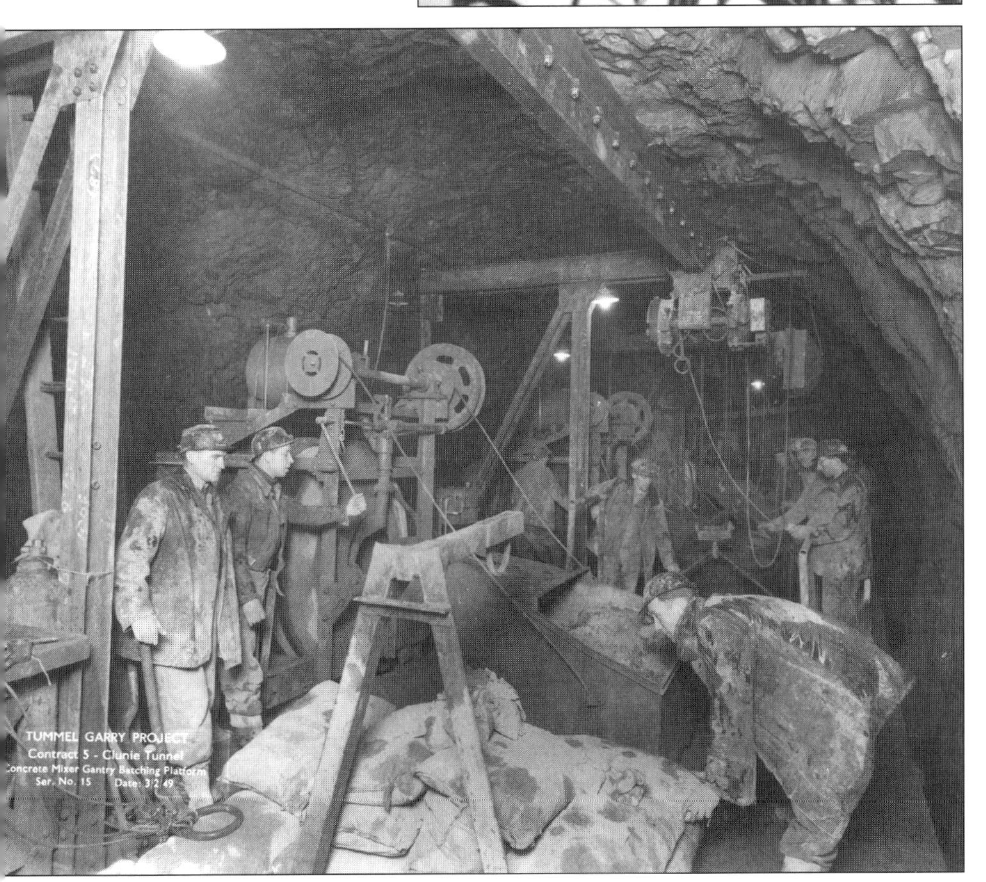

Plate 25.
Donald Macleod perched on the girders during the construction of the cement mixer housing at Glen Errohty (*Donald Macleod*).

Plate 26.
A concrete mixer and batching plant inside the Clunie tunnel on the Tummel-Garry project, February 1949 (*NOSHEB*).

Plate 27.
Opposite. Shuttering on the buttresses on the Meig dam, September 1953 (*NOSHEB*).

Plate 28.
Opposite. An early stage in the construction of the Meig dam, March 1952 (*NOSHEB*).

Plate 29.
The Meig Dam in 2001 (*author*).

wash. There was this guy from Caithness whose bed was next to mine in the camp and you could scrape the flakes of hard cement off his head. Off he would go in the morning with this crust on his head and he would come back with a new layer.'

It was standard practice to convey the hoppers with the fresh concrete from the batching plant out across the site to where they were needed on the blondin. 'The cable was nine-inch diameter steel, suspended 600-odd feet at the highest point from the river bed,' said Donald Macleod. 'You could move the masts holding up each end by 100-horsepower motors from side to side, angle them, and when they were installed we used to go out on what we called the bicycle, the machine that carried the skip for the concrete. This cable had to be cleaned because it was all grease and everything, and we used to go out on this bicycle 600 feet above the river. I had a good head for heights, I quite enjoyed it, you had a good view, and nobody would come and chase you. There was quite a big swing on it at times, when it was windy. The hopper was slung on chains and held a few ton of cement.'

Don Smith remembered the blondins at Trinafour with awe: 'The cables

Plate 30.
Laying the Allt na Fainich aqueduct on the Orrin scheme, August 1956 (*NOSHEB*).

had to be greased every day. The main cable was about three inches wide, rounded but flattened on the upper side, and it swayed. Blokes used to walk across this bloody thing, no safety belt. Two wires on either side, and two cables below for the thing carrying the cement. The blondins were 120 feet up, so in the middle these blokes were probably 200-250 feet high, just walking.'

When the workforce of Duncan Logan Ltd started to build the Meig dam, they had to make do with less sophisticated equipment, as Bill Mackenzie recalled: 'I put in the first shovelful of concrete in the Meig and the last one. I was in charge of the concrete squad – there were only four men. The firm was young then. Machinery and cranes were expensive. They were a wee bit short of material and it was all shovel. The rate of concreting was a five-ton crane delivering a five-ton skip every five minutes, and they were trying to speed that up to get on, as usual. We hadn't vibrators at first and we had to use the boot to settle the concrete. Now, the five ton was dumped and it was spread, until it was six to eight inches deep over the required area, by four men in five minutes.

'Maybe a whole day's pour would take sixteen hours and you had been at it all that time, five-ton batches coming every five minutes. They had two cranes but they could have done with another one as the furthest away part of the dam was beyond the reach of the crane and we had to shovel from man to man. That was work, I can tell you. It's a heavy job and the concrete's soft. As you're working you're maybe sinking up to your knees in it. A sore back all the time. We were supposed to be on twelve-hour shifts but once you started pouring you had to finish. It all depended on the size of the bay and there would be maybe wee snags along the way – it could be sixteen hours to get finished. Sometimes we were late in getting started, maybe 3 or 4 o'clock in the afternoon, and you would be on all night, working under floodlamps. Then you'd have an hour in bed and be back again in the morning. Once you started concreting it didn't matter what the weather was, although we had to stop if the frost was very severe. We didn't stop for rain. We had oilskins on and shovelling in oilskins and rubber boots was hard work. You couldn't complain; if you didn't like it, you knew what to do. It was an experience.

'Then we did the Orrin dam. That was a lot easier. More tackle, more men. I was still a ganger but I had moved on to fancier jobs, finishing and so on. I did all the crests on the spillways, on the Meig and the Orrin, and all the finishing jobs between. In the Orrin dam you can walk right through it, there's a passageway inside the dam. I built all the stairs inside. They were precast and dropped in.'

Orrin Project 17/8/1956
Orrin Tunnel and Allt na Fainich Aqueduct
Allt na Fainich Aqueduct

Plate 31.
A rock fault in the Orrin tunnel (*NOSHEB*).

Roy Macintyre remembers some of the men he worked with on the concreting of the Meig dam. 'Barney Malone was a ganger and was given the responsibility of vibrating the concrete, and he did it methodically and with great care. The vibrator was a heavy poker-like device to drive out air bubbles, equivalent to tamping. It was important not to over-vibrate the concrete as this would cause segregation [separation of the constituents]. It had to be done for just the right length of time. Now, Barney liked to go for his dram on a Friday night and most of the wages got blown. On this particular Saturday morning Barney had had a heavy night the night before and he was vibrating the concrete – and he couldn't have been feeling very well for he suddenly emptied his stomach and his false teeth into the concrete and, with the vibrator going, the whole lot disappeared straight into the mixture. Poor Barney was without teeth. He was fitted for a new set and, I remember all that summer, Barney would be getting ready to go to Inverness for his teeth – but you would see him on Monday morning and his mouth would still be sunken looking and he would say he hadn't got past the Achilty Hotel.'

Iain MacRae went to Invermoriston as the chargehand in a gang of shuttering joiners and his first impression of the site has remained with him: 'There was nothing there, nothing at all. They started by cutting into the rock on both sides of the river, away in maybe forty feet on one side and thirty feet on the other. Then that was all concreted to make a very big foundation: there must have been about an acre of concrete, what they called a solid core, on the bottom of the river. They diverted the river, put in coffer dams and worked on the inside of them. It was quite an education as we'd never seen this before.'

The shuttering joiners worked in squads of eight men to build, in effect, the moulds to receive the wet concrete. Shutters were made in the site workshop, from inch-and-an-eighth strips of dressed Swedish spruce. The joints between the strips had to be tight to stop the leakage of the frothy, watery cement run-off the men called 'fat'. Special skill was needed to build the oddly shaped shuttering that served as moulds for the heel of a shaft or a circular shaft mouth.

The joiners dealt with a 'lift' every day. This was the section of shuttering, seventy-two feet long and six feet high, that could be erected to make a mould and filled with concrete in one day – a day's pour. 'Keys' could be poured at the ends of a section to lock into the next pour, and copper and rubber seals placed to prevent leakage through the end joints. The shuttering was held to the face of the dam as it curved upward, by bolts run through reinforced steel joists or RSJs, known as 'soldiers' and often made from old railway lines. The sections of shuttering were raised and lowered into position by a crane or a derrick.

'The work wasn't desperately hard', recalled Iain MacRae, 'but the hours were long. I used to leave the house at six in the morning and not be home until nine at night. You had to be on the job at half past seven. It was thirty-two miles each way to and from Kirkhill, in one of Logan's lorries, a Ford Thames. An old fellow, Donny Urquhart, was the driver, they called him K. Don after the great racing driver, and he went like the mischief. He was one of three men who drove from this area: to collect the workers in the morning, they started from Strathpeffer, and went over Mulbuie, on the Black Isle, to Avoch and then to here and over the top by Culnakirk to Invermoriston. The weather didn't matter. At work, you never came in for rain or snow, you were outside all the time. We were given clothes, no hard hats – but everybody got oilskins. If you wanted a donkey jacket you had to buy it. It was drastic in the winter.

'We had some good laughs. There was a fellow, Eddie Chalmers, from Muir of Ord. He was a brilliant artist. Periodically we would get a brand new shutter and, as soon as Eddie saw it, he would do a drawing on it of

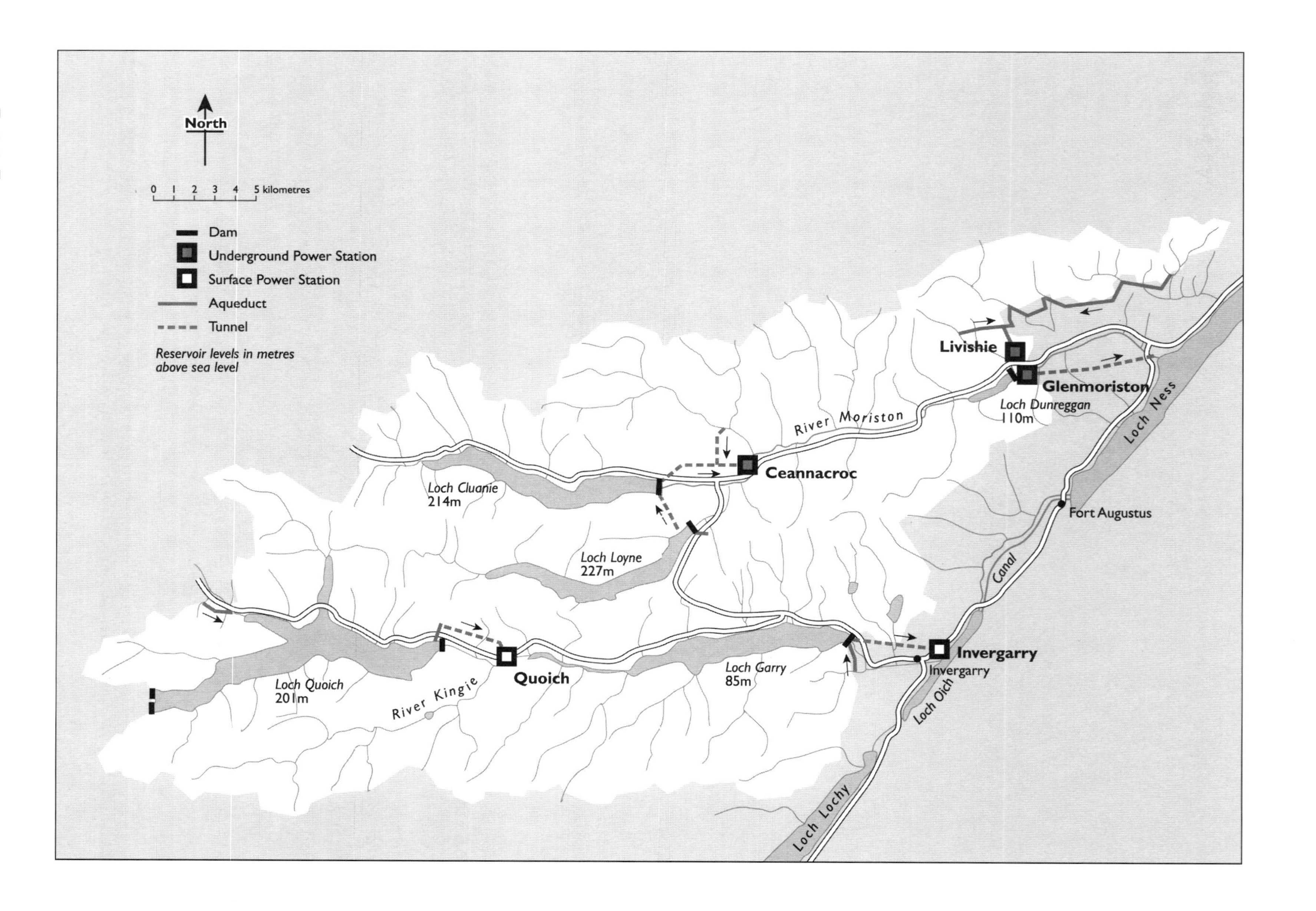

North
0 1 2 3 4 5 kilometres
Dam
Underground Power Station
Surface Power Station
Aqueduct
Tunnel
Reservoir levels in metres above sea level
Livishie
Glenmoriston
Loch Dunreggan 110m
Loch Ness
River Moriston
Ceannacroc
Fort Augustus
Loch Cluanie 214m
Loch Loyne 227m
Canal
Invergarry
Invergarry
Loch Oich
Loch Garry 85m
Quoich
Loch Quoich 201m
River Kingie
Loch Lochy

somebody, with a big joiner's pencil. Now the bloke in charge at this time was Willie Ingram. Eddie drew two pairs of Clydesdales pulling a plough, and Willie Ingram was the ploughman walking behind the horses. This was on the full length of the inside of the shutter. Walking along the furrow behind Willie came Tom Critchley the engineer with a level on his shoulder. It was brilliantly done. We used to keep our eye open for anybody coming but this time that didn't happen. Willie Ingram came round the head of the shutter while Eddie was busy doing this drawing. The shutter was up by this time, in position and bolted. Willie came in and said, "What the Dickens is that?" Then he says "Take that shutter off there." We had to unbolt it and take it off, and we left it leaning up against the office, and it was there for the duration. Willie was so pleased with it.'

Map 5.
Great Glen.
The Garry–Moriston schemes

Iain MacRae went on to work on the Orrin and Meig dams. The last job was the construction of fishladders on the Meig after the dam was complete. One of the joiners was an old-timer, Tommy Hossack, and he was complaining about the way they had set about the job. '"That's no the way to do it at all," he would say, "This is what we should be doing" and all this. This day, this fellow came in and started talking to us. He wore a fore-and-aft hat and plus fours. He asked how things were going and what we thought of the design. This got Tommy going, and Tommy told him it was no use. It turned out the newcomer was a Dutchman, the bloke who had designed the fishladder in the first place. He took Tommy's views all in good part. Afterwards he changed the design. It was very complicated, with stepped concrete, awkward shapes and corners, like building a stair in reverse.'

Patrick McGinley arrived in Inveraray after an all-night journey and went to work more or less at once on the Lochan Shira dam. Two ten-ton electric derricks had been lying idle for want of someone who could drive them. Patrick had picked up the skill of crane driving in England, after a workmate had said 'God, you've got a dirty owld job at that fitting. Why do you not start at the driving?', and had found it to his liking. Now, he and a mate were put on the derricks right away and worked a twelve-hour shift. 'That was it,' he recalled. 'We stayed there a good while.'

Later, he shifted to another site where his brother, Eddie, found him a position as a crane driver. 'A man asked me if I would like to go on a Jones revolving crane. A Jones mobile crane. I says, "I don't know much about that kind of work." "Ah," he says, "you'll be all right, you'll soon learn."'

The Jones crane had rubber wheels and a small diesel engine, and was used to lift and move the shuttering around the dam. The crane itself was lifted by the blondin from one construction bay to another, a short aerial trip during which the driver stayed aboard.

'It was kind of dangerous sometimes lifting shutters,' said Patrick. 'It was hard, you know, tight, they were stuck in the cement, and the joiners would be around hitting them. Some would come up no bother but there were some you needed to hit away at, from the scaffold outside, where there was a big drop down below. Once you got one shutter up you were all right, then you had a bit of protection for the wee crane. If you put too much weight on it, the wheels would come up, you know, the wheels would rise up, and that was it, the crane was away then. If you had a long jib on, you could lift a couple of ton, but this was a short jib, specially made for lifting the shutters, and you could give that bit extra power. If the shutter was tight to move, there would be maybe four- or five-ton pressure coming on it. A shutter itself would be about a ton at the very most.

'Everybody knew what they were doing, they knew the job. There would be no accidents or nothing. You would try and lift it first, and if you didn't move it, well, they would slacken the bolts off, do something, to make the shutter come away easy. There was no dragging of it. It had to come smooth.

'On a particular wall there would be three or four shutters. With the crane, you placed the shutters on the left, the right, behind you, like a square. You just moved the crane back and fore in that particular bay, whatever space it was. When you had all the shutters in place, and the joiners had them all tightened up, the blondin would come and lift the wee crane and land you in the next bay. I was still in the crane. I'd be shaking, and when I'd be put down I'd hook the blondin off. Depending on when that was done, whether it was time to start or not, you could go down. There was a big ladder. I was up maybe fifty, sixty, a hundred feet.

'I was young at the time. I could climb up no bother and I wasn't scared of heights. You wouldn't need to be scared. You could surely feel the crane swaying with the wind. But it was solid enough, it did the job it was meant to do – lifting the shutters.'

Patrick McGinley often stayed up in his eyrie above the dam for the whole of his twelve-hour shift. To avoid the climb down during the break for the midday meal, he would sometimes have a flask of tea and sandwiches up with him, and sit in the tiny cabin and read a paper. When he had to answer the call of nature, he just had to do it on the job.

Later, on the Glenmoriston scheme, he used a crane to drive steel piling. The ten-inch steel piles were positioned carefully and unshackled, and then the crane driver picked up the steam hammer and put that on top of the pile, where it was clamped in place. In the hard rock it could take some time to drive the steel down as much as possible of its twenty- or thirty-foot length. Ten or twelve strokes of the steam hammer might push it into the unyielding substrate by a miserable inch. The foreman decided when each pile was sufficiently embedded and, when the driving was complete, a man with an oxy-acetylene torch burned the pile tops to level them.

As was the custom, many workers moved from one scheme to another, according to their whim, and found a job where they were needed. A setback might set a man thinking 'To hell with this' and walking off. Patrick McGinley once left a skip of concrete hanging in the air at Pitlochry. He took over an old derrick temporarily from its operator, unaware that a running repair with a stick had been improvised to stop the cable from coming off the drum. The stick jammed. 'I took a scunner and left it there,' said Patrick. 'Left it altogether and went down to London. I don't know how they got it down, I wasn't interested anyway.'

In London, that time, Patrick met another McGinley he knew: 'Black McGinley they called him, Steam-iron Jim McGinley, no relation but he came from Donegal, he said we should go down to Hammersmith, to Wimpey's office, we might get something in the north. We went in. They knew this McGinley. He was well known. He was a lot older than I was. The man says would we fancy going up north, to Inverness? I didn't say nothing. Jim, he says aye, we surely will. The man gave us a voucher for the night train, some £20 each, to land in Pitlochry. "There'll be somebody to meet you there to take you to Clunie".'

A jeep met the men off the train and took them the few miles west to the site of the Clunie Dam, far enough from the last place for Patrick to escape any censure for leaving skips of cement in the air. The work at Clunie was to the young Donegal man's liking and he stayed 'a good while'. A foreman recognised his ability to 'jump on anything', to drive any type of crane or derrick, and offered him extra money to stay. 'I was spare driver then for a good while and I was getting the same money as if I'd been on the crane. I was on stand-by. If I was wanted or if somebody took a day off or got sick, I could jump up on any particular machine and that was me. The foreman says, "I'll miss you now if you go".'

The Hydro Board published two major new projects in 1948. In February the Highland papers carried the details of the Glascarnoch scheme, aimed at tapping the water resources in 345 square miles in central Ross-shire to produce 280 million units of power a year. Estimated to cost £8.5 million, the scheme would create four power stations and six dams, and involve the

Conon, Meig, Bran and Glascarnoch rivers and parts of the drainage areas of the Carron and the Broom. The level of Loch Fannich would be raised and new lochs would appear in Strath Vaich and in the long valley of the Glascarnoch beside the main road from Dingwall to Ullapool.

The Moriston-Garry scheme was published in the following May and, to the satisfaction of the editor of the *Inverness Courier*, involved no diversion of water from the Loch Ness catchment area. The West Highland Power Bill in 1928 and the Caledonian Power Bills in the late 1930s had all sought approval to develop the Moriston-Garry basin but had been defeated in Parliament; the older schemes had considered it necessary to divert water from Loch Quoich to the west, thus threatening the Ness with reduced flow and a danger of drought in summer. The new plans preserved the eastward flow of the river drainage and also promised to drown less land. In Glenmoriston only one house attached to the sawmill at the Blairie estate would be submerged, and the family who lived there told the *Courier* that they had no objection to moving and were looking forward to being nearer a place with more community life. Reservoirs would be created – in Glenmoriston and at Dundreggan. A large dam would be built at the east end of Loch Cluanie, extending the length of the loch by three miles. Here one lodge and one cottage would be submerged. Dams would also be constructed on Loch Loyne and Loch Quoich, and Loch Garry would be doubled in length. The raising of Loch Quoich would flood over a number of cottages and two lodges, of Loch Loyne one house that had stood empty for decades, and of Loch Garry a cottage and an empty mansion house. The whole scheme would cost almost £13 million.[13] In the summer of 1952, the design of the scheme was simplified to save steel and cement, both commodities in short supply, and a lower dam at Dundreggan was dispensed with.[14]

The Highlands received a further filip when the contracts for parts of the two schemes were awarded to a local contractor – Duncan Logan Ltd. This family firm had been founded in 1895 in the village of Muir of Ord at the west end of the Black Isle peninsula in Ross-shire. Duncan Logan, the founder, was a stone mason who acquired ownership of the Tarradale quarry and became the biggest house builder in the area. Those who remember him recall a man who worked hard and expected his employees to do the same.

The firm grew slowly for many years – the turnover in 1934 was £25,000[15]; the payroll in 1939 had fifty people – but Willie Logan, Duncan's son, took over the reins and proved to be a dynamic, almost buccaneering expansionist. Willie was born on Christmas Day 1913 and he joined his father's business after he left Dingwall Academy, the local secondary school, in 1932. His academic career had been fairly undistinguished but he had ambition and a remarkable mental facility with numbers and detail. He was also fired by a desire to work and succeed, and he drove the company to take on more and more contracts. His restless enthusiasm was almost an embarrassment to his fellow directors, reported one observer,[16] but it endeared him to his ever-growing workforce, many of whom had been his schoolmates, and made him a hero in Ross-shire. Laurie Donald was recommended to Willie by another engineer and was called to meet him in the Station Hotel in Inverness: 'He was an interesting person, full of go and vim. He could be quite hard if he thought he wasn't getting the cooperation he wanted but he stood by his men. They had great faith in him and he knew his foremen.'

'He would never ask anybody to do anything that he wouldn't do himself,' said Iain MacRae. 'Once some men [on the Glenmoriston scheme] left a set of levels out on a rock in the middle of the river. It had risen eight feet and this expensive equipment was in danger of being washed away. It was Willie himself that went down on a sling from a crane and picked it off.'

'Willie Logan and I got on all right,' said Bill Mackenzie. 'He would just say "Chance it, Mac". I had a job inside a tunnel on a shuttered face, where the concrete was going in by pump. It had to be well pumped to make it secure. This night he was there himself and the pump was going. I said, "Look, Mr Logan, the shutter's beginning to bulge." It was a terrible pressure, with 200 ton of cement inside it. Luckily enough, there was a bit of planking to hand and we were able to strengthen the shuttering. Logan said, "Gie it another pump, Mac." He was willing to take a chance.

'On the top of the Orrin dam there was a crane with a 120-foot jib. From the top of the dam to the riverbed, where a pipe had to be shifted, was 120 feet. It was only a five-ton crane and the pipe was six or seven ton. The jib was full out at 45°, and an overload along with that. The driver said no, no, he wasn't doing it. "Come out of there, lad," said Logan, "I'll do it myself." And he did, and you could hear the bolts screaming.'

Duncan Michael from Beauly left school in the summer of 1955 and, because his family already knew the Logans in neighbouring Muir of Ord, went to ask Willie for a job: 'He said, "Hello Duncan, yeah, yeah" and put me on a lorry. I was a favoured child and he put me on the various schemes but mostly I was at lower Glenmoriston, where I worked during four summers while studying engineering in Edinburgh. I commuted for a while in K. Don's lorry – he drove like mad. What I'll never forget is that at half past five in the morning in summer, it is magic in Beauly Square – the birds, the light – and this lorry comes in, you jump in the back, and the smell of that diesel matched against what air can be. Whenever I smell diesel I think of sitting in the back of that lorry.

'Willie Logan was a man of huge energy, a gambler, he looked for a problem and head-butted it. At Invermoriston we were tipping rock spoil into the Loch Ness and it's pretty deep. Once a driver on a huge machine was out there on the edge of the loch and lost his nerve. The tailrace works were difficult and, at times, dangerous. Willie – this great fat man – waddles over, gets in and drives it back. In corporate planning terms it was mad, staking the heart of the company on one machine, but he was like that.'

Willie's brother, Alastair, ran the quarry near Beauly that supplied the huge quantities of sand and aggregate needed on the schemes. 'Alastair was a completely different character from Willie,' recalled Roy Macintyre 'Willie never rested for two minutes, working at fever pitch, but Alastair was laid back.' If there was a knotty problem, Alastair was liable to disappear for a while and come back when it was likely to have been solved. But Alastair was also a man with ideas; he told Bill Mackenzie once how he would like to reclaim the head of the Cromarty Firth and turn the mudbanks into farmland.

Don West started driving for Logan after leaving Reed and Mallik at Mullardoch and stayed with the firm for eighteen years. 'We were driving sand from Beauly up to Corriekinloch at Inchnadamph, to Lairg, to Glascarnoch, to Quoich, all over. No matter who built the dams, Logan supplied the sand. He had thirty-two lorries on hire in addition to his own fleet of between thirty and forty. On the street in Beauly, you could hardly turn your head but a Logan vehicle of some sort would be passing.'

Willie Logan never seemed to rest and wise employees knew to look busy themselves whenever he was around. He had an abrupt, fast way of talking and would reel off an instruction or an order, leaving a man having to ask a colleague what had been said. 'I would ask the man in the office and he would say "Oh you're going to such and such a place", and I would be off around half of Scotland, delivering, loading, shifting – long hours,' said Don West.

'In the fourth summer I was working for him', said Sir Duncan Michael, 'Willie asked me what was I going to do. I could have had a job with him but he said, No, Duncan, there's a world out there. Off you go. I've been grateful to him ever since. He could have had me, I would have been loyal to him but he thought no, that's not fair. He could see an end to the schemes.'

Logan's fame rested on very Highland foundations. Teetotal and non-smoking, he was an elder in the Free Church and no one in the firm was allowed to work on the Sabbath. This observation of the holy day caused some controversy whenever there was a delay in completion of a job, but by and large he was admired for his steadfast adherence to principle. In the

Plate 32.
Willie Logan (third from the left) at the opening of the Orrin Dam (*courtesy Dingwall Museum*).

1950s the Highlands still cared about these traditions.

He lived in a large house in Dingwall – the beautifully laid-out garden was pointed out to visitors – but his main church was the Free Church in Muir of Ord, where he had gone as a boy. At the time of one of the dam contracts, on a Sunday, water started to leak into a site to be concreted on Monday morning; something had happened to one of the submersible pumps and the place flooded. They sent for the engineer who came out in his own car and collected one of Logan's vans at the yard to bring up a new pump. The engineer happened to pass the church as the congregation was assembling and, on Monday, he was called in to Willie Logan's office to explain this breach: "Man," Willie was reported as saying, "could you no have taken a machine that didn't have my name on it?"

Although some necessary jobs were done discreetly on the Sabbath, by and large Willie deserved the congratulations extended to him at the opening of the Orrin dam in April 1959 for 'having accomplished the huge construction in three years and ... without Sunday labour'. In the construction of the Orrin the men had set a record for laying 4,528 cubic

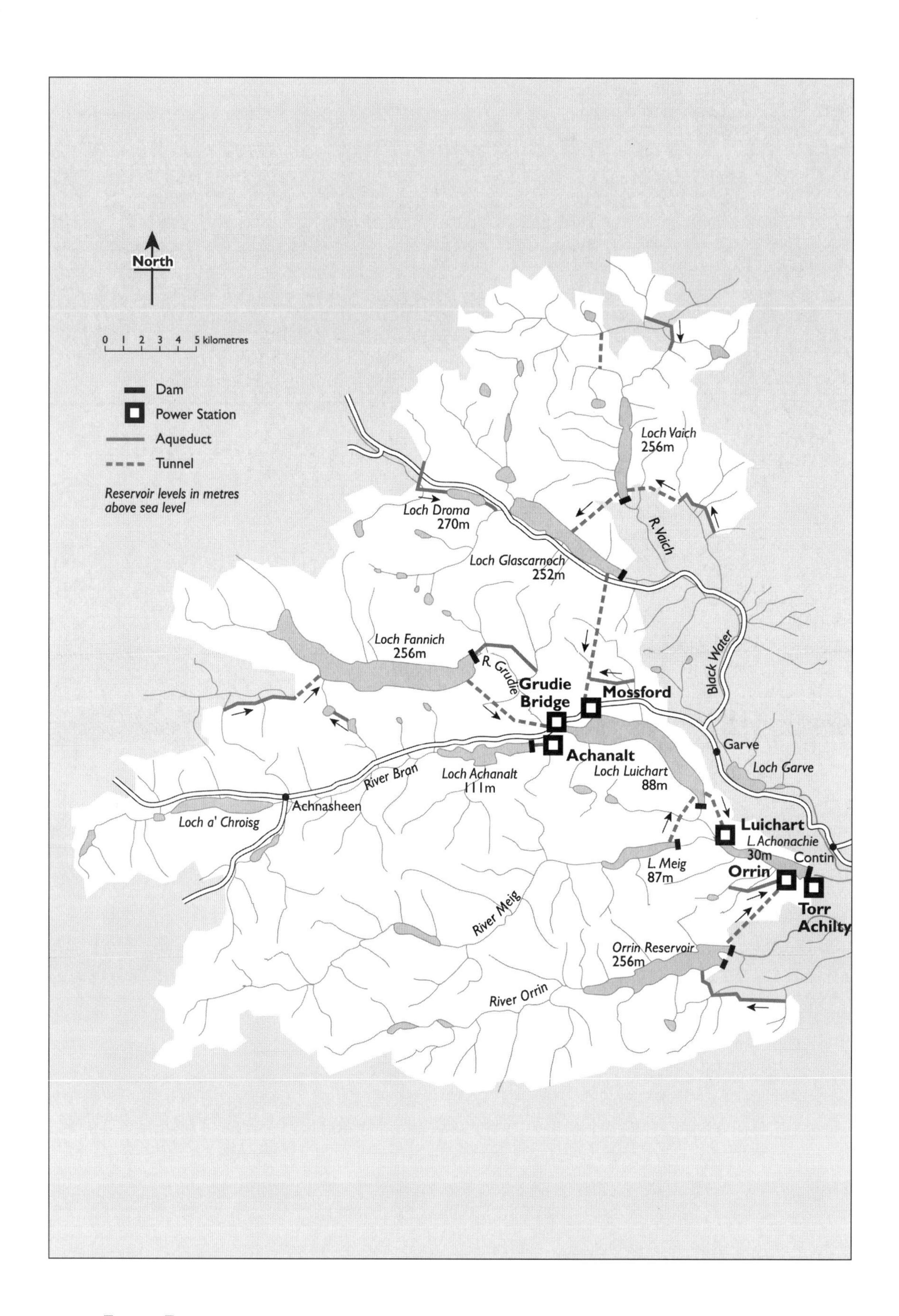
North
0 1 2 3 4 5 kilometres
Dam
Power Station
Aqueduct
Tunnel
Reservoir levels in metres above sea level
Loch Vaich
256m
R. Vaich
Loch Droma
270m
Loch Glascarnoch
252m
Loch Fannich
256m
R. Grudie
Grudie Bridge
Mossford
Black Water
Garve
Loch Garve
Achanalt
Loch Achanalt
111m
Loch Luichart
88m
River Bran
Achnasheen
Loch a' Chroisg
Luichart
L. Achonachie
30m
Contin
L. Meig
87m
Orrin
Torr Achilty
River Meig
Orrin Reservoir
256m
River Orrin

yards of concrete in five days; Willie had forbidden an attempt at a seven-day record because 'from the outset we wished to uphold our Highland heritage and programmed this contract on a "no Sunday" basis'. At the luncheon after the ceremony, further praise came Willie's way and he replied, saying that it would be vain of him to try to conceal his pride over the firm's achievements. Their earth-moving fleet, he said, was the best in Scotland, if not in Britain and in the next three weeks the entire operation would be moving to South Wales to fulfil an excavation contract.[17]

The first contract completed by Logan for the Hydro Board had been the construction of the small scheme on the Kerry Falls near Gairloch. 'I was with Halcrow the consulting engineers,' said Roy Osborne. 'Outside I would guess there were sixty men working at the peak. The concrete was mixed by hand and placed by wheelbarrow. There was a weir at the top of the glen on

Plate 33.
The pipeline following the Kerry river, Gairloch, from the reservoir to the power station, May 1951 (*NOSHEB*).

Map 6.
Opposite. Conon

the loch, and then a canalised river to the intake weir, and then a pipeline down to the power station. The top weir was just to give a little impounding to the loch. The lower weir was to create a catchment pond to guide the water to the pipeline. It took between two and three years to complete.'

The mainly local workers lived in ex-army wooden huts on a site close to the intake weir. This camp was run by Willie Logan's wife, Helen, setting a pattern for the years ahead. Roy Osborne and his fellow junior engineer lived in digs at Badachro: 'It was my first experience on a construction site and, coming from the sinful south, I was ready to work on Sunday, but the men would look the other way. They weren't rude but they wouldn't acknowledge me. It wasn't oppressive but it was quite noticeable.

'We started at 9 and finished at 5. I suppose we had weekends off but in the office we worked every Saturday morning. For entertainment we just went walking. Occasionally we went to the pub at Shieldaig. I met my wife in Gairloch. There were dances. The resident engineer, who smoked a pipe, went to this dance in the Gairloch Hotel, and a box of Swan Vestas went off in his pocket. It caused quite a commotion but no harm was done. The manager came up and said "I'm very sorry but this is a private dance", took us back to his rooms and poured us a drink.'

The Kerry Falls scheme opened in May 1952. Small it may have been but it had its share of problems. The workers cut the main telephone cable to Iceland, which happened to run down the glen, and a GPO engineer was sent from Dingwall at as high a speed as possible along the single-track road along Loch Maree. The excavation for the intake weir uncovered a pothole in the stream bed that proved to be some twenty feet deep and took several days to empty of gravel before it could be filled with concrete. The route of the pipeline down the glen was altered to save some trees which later blew down anyway. The sections for the pipeline were made in Motherwell and welded *in situ*, after being lined with bitumen in a machine designed for the job by Willie Logan himself.

Roy Osborne and a colleague walked down through the completed pipeline from the weir to the power station, crouching in the four-foot diameter space. When they were inside, Roy thought he could hear the rushing of water. At the end, the other man said he too had heard it. Neither had felt able to mention their apprehension that they were about to be caught in a flood.

In the decade after 1949, the Hydro Board awarded contracts worth over £8 million to Logan. A quarter of this sum formed the firm's wage bill, a measure of the contractor's importance to the Highland economy.[18] Willie Logan admitted that the first contracts from the Hydro-Electric Board had given him the experience and the self-confidence to expand the firm; the same self-confidence, he asserted, had been 'also released in hundreds of my fellow countrymen'.[19] It was said that Tom Johnston had originally invited Duncan Logan Ltd and R. J. MacLeod, another contractor with a strong Highland connection, to tender for Board projects to counteract what was felt to be overpriced bids from the big southern contractors: Roy Osborne remembered handling tenders in the offices of Halcrow the consulting engineering firm and finding that bids from Duncan Logan Ltd and R. J. MacLeod were between 40 and 60 per cent less than the others.

The competitive pricing paid off. Throughout the early 1960s, the Logan operation included contracts for the Gartcosh strip mill, the Dunbar cement works, the new Ness Bridge, a NATO fuel dump on Loch Ewe, the Corpach pulp mill, the Tay road bridge, and innumerable roads and housing schemes. As the head of the enterprise, Willie's bulky, square-cut, bald-headed figure often featured in the Scottish press, usually as the man from the Highlands who was scooping all the big boys from the south for lucrative and prestigious contracts.[20]

Roderick John McLeod, or 'RJ' as he was almost universally known, was actually born in Canada but he lived for a while at Elphin in Sutherland, and his mother ran a café in Ullapool. After his Highland childhood he worked for the Aberdeen contractor William Tawse Ltd before setting up on his own, with a headquarters in Glasgow. 'RJ had been the resident engineer on the road along Loch Lomond and through Glencoe before the War', said Roy Macintyre, 'and later he started his own firm. He was probably on a par with Logan although a bit smaller.'[21]

The Hydro Board's first projects to be completed were two small, local ones – at Morar, and at Nostie Bridge near the Kyle of Lochalsh. They came into operation on Tuesday 21 December 1948 and made electricity available to the country between Mallaig and Lochailort, and to a large area of Wester Ross and a part of Skye. Catherine Mackenzie, the widow of a crofter, performed the opening ceremony by throwing a control switch on the turbine at Morar. '*Gun tigeadh solus agus neart dealan dhionnsuidh gach croit*' ('Let light and power come to the crofts') she was reported to have said in Gaelic, and she added 'Electricity means new hope for the Highlands. Thank God for it!' Tom Johnston was delighted and told the press it was an end to the old argument that oil lamps were good enough for the crofters.[22]

The large schemes took several years to complete, bedevilled as they still were by labour problems, shortages and bad weather. Two years after Mrs

Johnston had fired the shot by the River Glass, work was progressing in Glen Cannich and Glen Affric at a rate that allowed prediction of completion by the end of 1950. (It was, in fact, to be another three years before the Duke of Edinburgh was to perform the official opening.) The original intention to build two power stations had been changed in favour of only one, at Fasnakyle. In June 1949, the great dam at Mullardoch, the largest the Board was to build, and the Benevean dam were slowly being assembled; two-thirds of the tunnel between Mullardoch and Benevean had been excavated, and the other main tunnel, from Benevean to Fasnakyle, was 80 per cent complete.[23]

'First week in December 1948, I started at Mullardoch dam as an electrician,' said Don Smith. 'I took a bus from Inverness to Beauly, another bus from Beauly up to Cannich, and hung about there waiting for a lorry going over the hill. I sat in the front with the driver. First week in December. It was snow and sleet, and it was misty, and I saw that camp with the smoke coming up from the cook house, and all the old Nissen huts, and thought what the bloody hell am I doing here. I'd never seen landscape like this before, at least not to live in it. We'd sailed up the west coast [in the Royal Navy] and been in the sea lochs, and we'd seen the hills, but I'd never seen anything like Mullardoch – the hills closing you down.

'The road up was single track, not finished, tarred in places. Anyway I got in, got to the camp, got a bed and a locker in a hut, and there were twenty-two or twenty-four beds in that hut. From the camp up to the job was about half a mile, you had to walk up in the mornings, up a dirt road, it was a bit coorse. I was there from December until the following November.'

One of Don's first jobs was wiring the crane and derricks on a Bailey bridge, built on stilts across the back of where the dam was going up. The wires were fed from a main cable running across the bridge, from a big iron junction box. This was not an easy task: as was common in the early days of the dam building, much of the material was old, ex-War Department stock, bought by the contractors on the cheap. The main cable, four inches in diameter, was stiff and required several pairs of hands before it could be made to feed into the junction box.

One of the characters Don got to know was a short Glaswegian called Tommy Cowie, 'the best wee public-works man I ever saw'. Tommy asked Don where he had worked before. 'Balfour Beatty.' 'Who was you with at Balfours?' 'My boss was Ernie Wells.' 'Ernie Wells,' cried Tommy, 'I knew that bastard twenty years ago at Spean Bridge and Tummel Bridge on the schemes, a Newcastle man. He was the engineer with Balfour Beatty, and then he got moved to Loch Sloy when that started.' Don learned some useful things from Tommy, such as the Glaswegian's habit of finding his own jobs.

Plate 34.
Tom Canning, from Ireland, checking a Bugee pump in the Mullardoch tunnel (*Don West*).

Plate 35.
The completed Mullardoch dam in winter (*Don West*).

Plate 36.
Don West at the wheel of a Dodge truck (*Don West*).

Plate 37.
Hans (surname unknown), a German, in an old Chevvy at Mullardoch (*Don West*).

'He used to come into the workshop, pick his bag up, leave his haversack with his lunch, and he was away,' Don recalled. 'I would say "Where are you working?" "Oh," he would say – and the chargehand used to stand there at the bench with his hammer, "I just look for jobs. They say where are you, Tommy, and I say I'm at the crusher or I'm at the sand plant, or over at the tunnel, or so-and-so. You see?" He never had a job given to him, he used to skive off, he did work, he knew what work needed doing and he'd move around and get what he wanted from the store.'

Don took a cue from Tommy and, after about two months, became a roving electrician: 'I just looked for jobs. The crane was nearly ready, so I went in and wired it. The fitters got to know you; they'd say "Hey, Donnie, there's a wee job here". There were four electricians in all and a couple of labourers but, at periods, I was the only electrician, with a couple of mates. Tommy Cowie left. He went down to the Tummel Bridge scheme but he wrote to me and said why don't you come down here. I was with Cochrane, on two shillings and elevenpence-ha'penny per hour at first, but then we had a rise to three shillings and a penny. Tommy said, pack that job in, come down here and work with me, I'm with Carmichael – far more money, more hours and a bonus.'

When he was working on the Trinafour dam, Don shared a cubicle with a young Irish lad called MacMullan. They worked in the tunnel on opposite shifts: 'I did twelve hours, he did the other twelve. Now there was always work to do. After a fortnight or three weeks I says to him, "What the bloody hell's this? I'm doing all the work". "Ah, you can do it," he says. He was a chancer, a nice mate, a good lad, but he wasn't an electrician. He had been a mate at Sloy. "Ah well," I says, "if you start to do a bit more, I'll leave instructions for you. You can soon pick up the job." Anyway, I was on days and doing all the bloody work – and then there would be half a salmon left for me. MacMullan was going off poaching with gelignite and a slow fuse. Now the man in charge of supplying explosives was scarred all around his face, the result of an accident with fuses in the pits. Anyway, him and MacMullan got on well, and I used to get my bit of salmon. I used to cook it on a wee electric fire in the cubicle.'

At Tummel Bridge, the electricians had to look after the 120-volt Oldham batteries used to power the electric locomotives in the tunnel. Heavy awkward things, the batteries needed recharging and topping up with distilled water every day. A charged battery lasted for eight hours and recharging took as long – on top of that was the lugging of the things, as long and as wide as a table, on and off the locos. Distilled water in quantity was not an easy commodity to come by in the Perthshire glens and a distillation apparatus was acquired to provide a continuous supply. This apparatus resembled an outboard motor attached to a container and produced distilled water at the rate of one drop per second – it took twelve hours to fill a carboy. Once it went missing, said Don Smith; it had been 'borrowed' to assist in making moonshine whisky.

Iain Macmaster worked as an electrician on the Lochay scheme. 'I looked after one site that branched out into two. From that site you could go right through the hill and come out on the other side once the tunnels had been made. Initially there was an electrician on every site, and sometimes a day-shift and a night-shift man, and I met up with my opposite number at the change over and we had a wee briefing. I did most of my work at the weekends – you got a clear run, after the tunnelling men went off shift at midday on Saturday and sometimes it would be Sunday night before they came on. As they tunnelled, we followed, extending the lights and the fans, and the best opportunity to do that was at the weekend. In the tunnel there would just be myself and a mate, and fitters who were taking the same opportunity to put in air ducts.

'You could say the standards were rough in comparison with today's but advances were being made all the time and we had hard hats. Usually the engineer and myself had electric miner's lamps. With the tunnellers on bonus, if they had power, light and fans, you were a good boy – if they didn't have any one of these, comment would be coming your way. I used to go in there on shift at 8 o'clock on a Monday morning, coming off at 8 on Monday

Plate 38.

Engineering and office staff of John Cochrane & Sons at the office building at Fasnakyle, December 1948. Mairi Stewart is third from the left in the front row, and Sybil Davidson is second from the right (*Sybil Davidson*).

night, going on at 8 on Tuesday morning and coming off at 6 on Saturday morning. Today that would not be allowed. By that time on the scheme the other electrician would be down about the camp to cover other sites, and the tunnel boss would say "I need you tonight because we're concreting, and if that power goes off I've got thirty men idle". There were plenty of blankets up at the tunnel and it was a case of "we'll get you if we need you". This happened on a fairly regular basis. At the head of the tunnel, there was a wee Red Cross room, hence the blanket, and some workshops, where I would sleep. Very often I wouldn't be called on to do anything.'

Archie Chisholm left school when he was fifteen and after working with his gamekeeper father for a time as a gillie decided, in October 1948, to try for a job on the Glen Affric scheme. 'I was interested in engineering,' he said. 'I knew one of the chaps there, the head engineer, Malcolm MacNeil, and I wrote to ask him if he would take me on as an apprentice. I started working as an apprentice fitter, gopher, making tea, and washing engine components with diesel or petrol, just a general run-around. That was a real experience for me, a naïve glen boy thrust into this very cosmopolitan environment. There were people from all over the world there. I was still staying at home and taking the bus up every day. They tried to employ as many local people as they could, [but] there were Poles, Lithuanians, Irish, and Canadians who had married and stayed on after the War (they had come over in the Forestry Corps), and also Maltese – just a tremendous mixture of people. Of course there was the usual overspill from Glasgow of engineers and marine engineers. I got a wealth of experience from them. I was lucky to serve my time with Davie Main from Banff who had been a first engineer on the *Stirling Castle* – what he taught me was tremendous. Of course, Cannich camp, built on a wartime basis by James Laidlaw from Glasgow, was over-lavish for that time – there was a barber shop, a shop, a club, a cinema and dance hall. Everything was on a grandiose scale. There were some 2,000 men in Cannich and Mullardoch, with some real characters among them. The five years I spent there was the best education any boy could get. It made you streetwise and worldly wise.

'On the first day, I was told to report to the workshop, a huge ex-RAF double hangar where they had their plant, with railway lines running in, and a big Motor Transport place where they did the lorries. The uniform on the scheme was a donkey jacket and a pair of welly boots, and I was told to go to the store to pick these up. A white donkey jacket, well, off-white, in a blanket material, was for ordinary workers, the staff got navy blue ones. Where they got them from I don't know, but I couldn't get one to fit me – it

Plate 39.
Left. Archie Chisholm beside a contractor's lorry at Cannich (*Archie Chisholm*).

Plate 40.
Right. A group of apprentices on a jeep at the Cannich work camp (*Archie Chisholm*).

was too long, but that didn't matter. It was all ex-WD stuff, even the plant – Cochrane must have gone to the yards and bought it all up. For going over the rough terrain they had Willys jeeps, still with the [US Army] star on the bonnet. Cochrane's colours were maroon and gold – but you could still see the star through them. All the engineers had a jeep. The firm also used Chevrolet four-wheel-drive lorries, ideal for the job. The narrow roads were difficult, tarmacadamed but narrow and dangerous.

'There were always pranks. Especially in the smiddy. Kevin Gillespie from Donegal and Willie Miller from the Glasgow shipyards were both characters. Kevin and a Canadian, Elmer Turner, were blacksmiths and worked together on two anvils. Kevin used to fry steak, bacon and eggs on a shovel. There was another man from Glasgow, Tommy Lennon, who was a piper. He was teaching Kevin to play the chanter but Kevin didn't seem to have the puff to keep the instrument going. They rigged a wee air pipe from the smiddy bellows to the chanter and then you would hear "The Boys of Wexford" playing away. We often had to make rails up for the tunnel, and level crossings. It was all riveted together. You would be working away quite happily and all of a sudden you would feel something burning in your pocket: somebody had slipped in a hot rivet. I was young and full of pranks myself then, and it was sort of tit for tat. As the only apprentice in the shop, I was always the tea boy. On the day I came I was presented with this half-five-gallon drum with a wire on it: this was for the tea. I couldn't believe this at first. I had to fill it with water, take it through to the smiddy fire, boil it and put the tea in. I hadn't much idea how to make tea and put in a whole

Map 7.
Shin

packet, and poured a tin of condensed milk in on top – it was like tar when it came back. I got a right rollicking. Every Christmas they put round a tin for me, and I came away with a big collection; they were very generous. There were about thirty men altogether in the shop. They repaired all the plant, did all the welding, repairing pneumatic tools and compressors.

'For my first two years I was working under an Aberdonian, Jimmy Thompson, who was in his fifties at the time. We were working, just the two of us, late one Friday on a rush job. When we had finished Jimmy decided to wash his boilersuit and sent me to the petrol pump to get some red petrol (you simply helped yourself to this). I put a copious amount into a tub used for cleaning engine parts, and Jimmy duly got down to his washing. When he had finished he hung the boiler suit up to dry, looking well pleased with himself, and then proceeded to roll a cigarette. He turned to me with his usual "Right, ma loon" and struck a match. There was a mighty flash and up went the petrol in flames to the roof of the hangar. Apart from his bushy eyebrows getting a singeing, he was okay – but worse was to come.

'We both raced out of the inferno to get help but, unbeknownst to us, an electrician had just been renewing the spotlamp bulb above the workshop door and had left the extension ladder leaning there. As we pulled back the workshop door, the ladder came crashing down on Jimmy's head, knocking him cold. Here I was – the place on fire and the only help lying unconscious with a gashed head. I raced to the garage about a hundred yards away and luckily some of the mechanics were still working. We mustered as many foam extinguishers as we could lay our hands on and managed to put the fire out but not before quite a bit of damage had been done. By this time Jimmy had come round and was taken to the sick bay to have some stitches.'

Archie formed a close friendship with the Canadian blacksmith, Elmer Turner, who had come to Scotland during the War and had married and settled in Glen Urquhart. 'He was a real craftsman, and a man of the wide open spaces, a big strong man. I loved to listen to his stories. He restored an old flintlock gun, making all the parts, and got it working. He made the lead shot out of molten lead on the forge, and shot crows. He used to walk at night back over to his home in Balnain from Cannich – a fair distance. One night he was going home and he shot this hind on the way. He gralloched it, cut it up, stuck it over his head and carried it all the way home. You can imagine the state of his face with the blood coming off the carcass. He stopped to take a rest and somebody came along the road, and thought he had been in an accident. Elmer explained what he'd done and added "I haven't been idle, have I?"'

March 1951 saw publication of more hydro-electric schemes for the

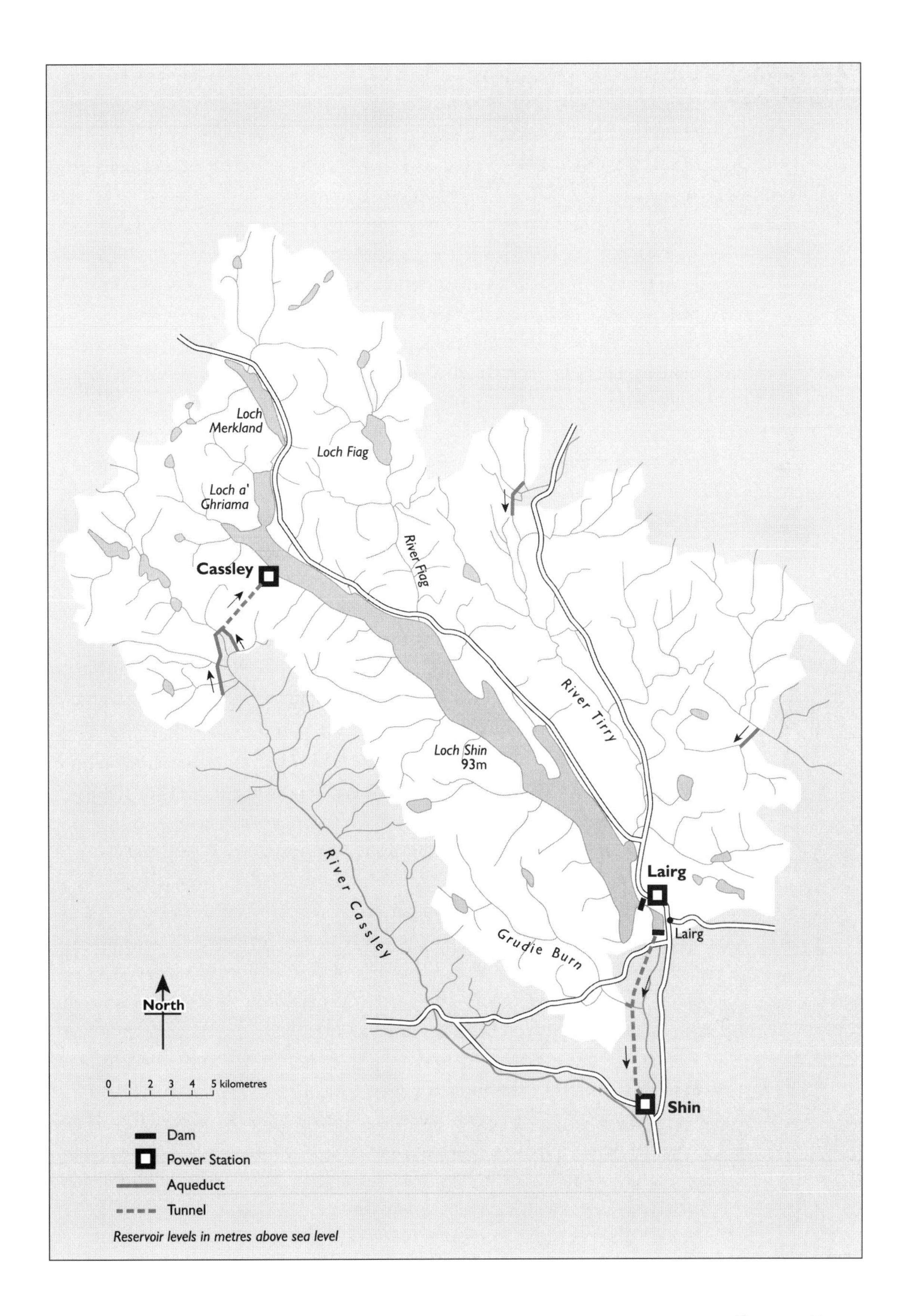
Loch Merkland
Loch Fiag
Loch a' Ghriama
Cassley
River Fiag
River Tirry
Loch Shin 93m
River Cassley
Lairg
Lairg
Grudie Burn
Shin
North
0 1 2 3 4 5 kilometres
Dam
Power Station
Aqueduct
Tunnel
Reservoir levels in metres above sea level

Highlands. The £8 million plan to build a dam on Loch Shin, to be the Board's most northerly hydro-electric project, was revealed at the beginning of the month; and some ten days later the third phase of the Conon basin schemes, the construction of a £3.5 million dam on the Orrin river, was announced.[24]

Power from Glen Affric flowed into the Highland grid for the first time on 19 July 1951 when Sir Hugh Mackenzie switched on the first of the three 22,000 kW turbines in the Fasnakyle power station. The completion of the Benevean dam and successful preliminary tests made possible the historic moment, although its significance was tempered somewhat by the fact that the dam at Mullardoch was still to be finished and the turbines could not yet be brought into full operation.[25]

It was over a year later, in October 1952, that the Duke of Edinburgh performed the final opening ceremony of the Glen Affric scheme. Under a sky threatening drizzle, the Duke drove from Balmoral to Fasnakyle and arrived too early, a little to the discomfiture of the officials. He walked through the cheering crowd, entered the camp dining room through the wrong door and approached the Lord Lieutenant of Inverness-shire and Tom Johnston from behind.[26]

Don West worked on an addition to the Glen Affric scheme about two years after the official opening: 'John Cochrane and Sons had done the main job, and then as an afterthought by the Hydro Board, which was pretty clever really, they contracted Reed and Mallik to install a turbine in the tunnel between Loch Mullardoch and Loch Benevean to generate power from the water flowing down. [After the inlet to the tunnel had been sealed by a concrete wedge and the water had drained out] they sank a shaft ninety-six feet straight down and about thirty feet in diameter to build the turbine. It was a tremendous job. They bored down through the rock ten to fifteen feet at a time, took out the spoil in buckets with derricks and tipped it over the edge of the dam. It took nearly two years. The vertical section probably took the best part of a year, nine months anyway. The buckets were loaded by hand. Once they reached the bottom they installed a lift – it's still there – and the equipment for the turbine was taken down. The firm of Gilkes and Gordon did the pumps, and the turbine was by English Electric. The turbine sits on a bed of concrete above the water tunnel.'

The turbine successfully installed, the next job was to remove the wedge from the tunnel inlet to allow the water to flow once more. The wedge, a huge plug of concrete, had been sealed with blue clay from Invergordon and was held *in situ* by the pressure of the water in the newly enlarged Loch Mullardoch. 'In the two years on the turbine job', continued Don West,

Plate 41. Jimmy Speed, from Invergordon, in the Mullardoch tunnel during the waterproofing of the concrete before installation of a generator (*Don West*).

'everything had gathered at the wedge and, when the turbine was ready, the wedge would not budge. A team of divers who were there to hook it to a crane tried for two months, working in the dark, getting nowhere. We had constructed a raft for them from four barges joined with twelve-by-twelve timbers, with a platform for their compressors, winches, pumps and so on. The only thing to do was to empty the loch. The engineer started to undo the bolts on a six-foot diameter valve at the bottom of the dam. There were over sixty one-and-a-half-inch bolts. Soon the water was gushing out, up in the air like a fountain. You can imagine the pressure – there was ten miles of water behind the dam, remember – and eventually he couldn't turn the bolts any more. Some more equipment was brought up from Cannich and they finally got the thing undone. Did that water not pump out of that dam? It went 100 feet before it hit the deck. There wasn't an awful flood as the outflow was restricted and the catchment area could cope, but it took weeks for the dam to empty. When it did, and we saw the wedge, we realised the divers would never have cleared it – there were trees, sand, silt, even dump trucks in there.'

The team of divers who tried in vain to free the wedge were an interesting microcosm of the multinational nature of the construction world – headed by Tich Fraser,[27] they comprised two men from Liverpool, one from Harris and one from Jamaica. 'Johnny Dailey was the Jamaican,' said Don West. 'An out and out gentleman. I used to drive the men to Inverness on Saturday, leave some of them, and pick them up on Sunday night to take them back to the camp. I said to Johnny this Sunday, come on down for a run, man, to my home in Struy. Johnny was learning the guitar at the time and I put on this Slim Whitman record, he was very popular then. This was great stuff. Then Johnny went away to get married and, while he was away, we decided to gather some money for him for a present. We must have collected about £35, there weren't that many of us on the turbine job. Johnny came back and showed us the wedding photos – he looked like a film star. Then we gave him the money, and the poor man was crying, he'd never got anything like that in his life. He was so taken.'

The next big scheme was published in 1953. Work had begun on the first phase of the Breadalbane scheme in 1951, with the start of construction of the Lawers dam on Lochan na Lairige on the south-west side of Ben Lawers to feed the Finlarig power station on Loch Tay, an arrangement that is incidentally the one with the highest head (1,348 feet) of all the Board schemes; but it was now proposed to add to the scheme by constructing in the Perthshire glens a further five dams and six power stations with associated tunnels and aqueducts, in a complex arrangement that sprawled across three watersheds. The overall cost was estimated at £15 million and the output at 304 million units of electricity per year.[28]

In 1953, Ronald Birse and his family arrived in the village of St Fillans. Growing tired of commuting between Biggar and Edinburgh, where he worked as an engineer in the capital's water department, he had applied for a job with Sir Murdoch MacDonald and Partners, the consulting engineers overseeing the construction of the St Fillans section of the Breadalbane scheme, at the east end of Loch Earn.

The quiet little village found itself suddenly at the centre of a great deal of activity, possibly the biggest thing to happen locally since the coming of the Caledonian Railway in 1901. 'We must have doubled or quadrupled the population in St Fillans and Comrie, where most of the engineers were billeted or lived,' said Professor Birse. 'There was a camp for most of the labourers and tradesmen, in a place called Twenty-shilling Wood. I don't

Plate 42.
Assembling a turbine in the generating hall at the St Fillans power station (*Perth Museum. Copyright: Louis Flood*).

Plate 43.
The Gaur dam nearing completion, and a shuttering joiner showing a good sense of balance (*Perth Museum. Copyright: Louis Flood*).

know why it had that name but it's still called that – there's a caravan site there now. Just outside Comrie. The camp probably had, I'm guessing wildly, somewhere between five hundred and a thousand men, living in wooden huts.

'We established contact early on with villagers because my wife and I had one child and were expecting another one, and we stayed in a caravan in the garden of a lady's house in Comrie for a few weeks until we found a house to let in St Fillans. Through this lady we met other locals and we certainly got to know quite a few. The residents of St Fillans before we arrived must have been hardly 100 – it was really a tiny place. It had a country dance society and there were two, if not three, hotels. It had a little tourist trade, but so much less than it is now. It's a watersports centre now, with hundreds of boats. There would have been a few boats then but nothing like now.

'We were quite isolated up there, though not as much perhaps as the engineers and workers on the remoter schemes. I suppose it was the nearest scheme to Edinburgh. We were thrown on our own resources for entertainment. My oldest daughter was born in 1952, so she was two when we went up there, and about four and a half when we left, and by that time she had acquired a remarkable vocabulary through playing with some of the other kids round about. The father of some of them was the tunnel boss, and we were amazed at the colourful language our daughter came out with.

'There were probably eight or ten engineers and their families in St Fillans, a similar number in Comrie, and a few in Crieff. I can remember one engineer, a foreign chap, he stayed in a cottage halfway between St Fillans and Comrie. To some extent we were all doing the same job but on different parts of the scheme. There was a superintending engineer and three resident engineers who had overall charge of a big area, and there were section engineers, of whom I was one, who had direct, hands-on responsibility for a particular part, in my case the underground power station. There were several section engineers on the dam, I suppose, and some on the tunnels. As resident engineers, employed by the consulting firm, we didn't have much direct contact with the labour force. In fact, we were not supposed to interfere with the labour force at all. If we had any observations to make, we were supposed to go to the contractor's engineers.

'Quite a number of the consulting engineer staff were English, because Sir Murdoch Macdonald had his office in London. Sir Murdoch had made his reputation in Egypt with the Aswan Dam. He was a venerable old chap by the time I met him – he came up two or three times to the scheme. The engineers were mostly young men, in their late twenties or early thirties, and for some it would have been their first big engineering job. So we had the two sides – the resident engineering staff and the contractor's engineering staff. The Hydro Board didn't supervise schemes directly themselves, they employed consulting engineers for each scheme.

'We were overseeing the contractors. We didn't have responsibility for the construction but we had responsibility to see the work was done according to the drawings and specifications. The contractor's engineers had the significantly more difficult job. We were pampered in a way: emergencies apart, we worked by and large 9 to 5, or 8.30 to 6, but the contractor's engineers, if they were working shifts in the tunnels or other sites, had to be on duty or on call all the time.

'We also did surveying and some of the setting out. The early work involved establishing base lines because we were working in country that hadn't been surveyed in detail. You had to do your own kind of local ordnance survey. I remember we set out a base line with very accurate theodolites and steel tapes, and tried to maintain an accuracy of a millimetre in 100 metres. I can't remember the actual accuracy we achieved. We had to project from that baseline, which was probably of the order of half a mile long, ten miles in one direction and five miles in another with as much accuracy as we could achieve. We tramped across the hills a fair bit. I was a lot younger and a lot fitter then.

'We got on well with the contractor's engineers. Occasionally they might be wary of us but if they were trying to skimp or rush anything they would know what they could get away with, broadly. They could always get away with something but this is what factors of safety are all about. There was a tremendous responsibility on the shoulders of the engineers. They had vast amounts of money under their control, and they were responsible for getting the job done within the time and within budget. As consulting engineers, we had none of that pressure, except that we shouldn't unreasonably delay things or make things difficult for them. They would have had a real gripe if we had been unusually pernickety.

'The day-to-day work on the scheme has to be carefully planned. Nobody is going to do anything down the line unless they've got clear instructions, so the guys wielding the pick and shovel or drilling to place the explosive have to know exactly what's expected of them. That information comes from the drawings and documents first of all, and usually the consulting engineer draws up a programme for the works, covering the whole project, in effect a chart outlining the parts to be under construction at various times over the total construction period, which might be three to five years. The contractor would then have to plan it all in much more detail. He had to do it for each part of the work – which gang was doing what where when, etc. It had to be all organised down to the last guy, the last worker.'

Roy Macintyre, as an engineer with Willie Logan on the Meig dam, remembers his dealings with the men from Sir Alexander Gibbs and Partners, the consulting engineers overseeing his part of the job. 'Each lift had to be got ready for pouring the concrete, a lengthy process in those days. You had to send a note to the consulting engineers to say Lift Number such and such was ready. The consulting engineer would come down to check all the setting out. If you got a sheet back that said it was a quarter of an inch wrong in this direction or an inch wrong in another, or the levels would need resetting, you'd go back and either agree or disagree with him. Then there would be a meeting to resolve the differences and you got the official go-ahead. It was very formal. All the paperwork had to be signed off. I think in later years it became less formal. Sir Alexander Gibbs and Partners were the *crème de la crème* of consulting engineers, or so we were told, and we did everything strictly according to the book. And then when the concrete was poured you had to be on hand. The degree of precision was probably far greater than was strictly necessary but it was a very good discipline.

'The consulting engineers also had inspectors, generally more elderly tradesmen, who were meant to do sub-tests on the work, checking the quality of the concrete by testing cubes. As I was a young fellow I was tasked with looking after the making of the concrete cubes. Sometimes, human nature being what it is, the contractor was always making damn sure his cubes passed the test. Some of the inspectors were very diligent.'

Concrete is not very exciting to most people. Redolent of urban utilitarianism, it symbolises the very antithesis of beauty. Yet its properties make it an invaluable building material and most of the dams were built with its grey masses, by the ton.[29] As an employee of the consulting engineers on the Breadalbane scheme, one of Ronald Birse's main tasks was making concrete cubes by the thousand and testing their quality.

'We took samples and made six-inch cubes,' he explained. 'We cured them, kept them in conditions of temperature and humidity similar to those of the dam, tunnel lining or wherever the concrete was, and then crushed them, destructively tested them in a big machine that put a force of seventy to eighty tons on the cubes.' The factor of safety the engineers were seeking was concrete strong enough to withstand a force of probably 3,000 pounds per square inch at twenty-eight days.

The weather affected the setting, the curing, of concrete. 'If it is not treated decently in its first seven days, if it is not protected from extremely

low temperatures, if it's allowed to freeze, and concrete can actually freeze in its first days when it has a lot of water in it, that destroys it completely, and it'll never get any strength at all,' said Professor Birse. 'Even if it doesn't freeze, low temperature will mean that it takes longer to achieve the required strength. In winter, you have to allow for the fact it gains strength more slowly. You daren't put loads on it, significant loads, at an early age. In summer, you could put a load on it in certain circumstances after three days, in winter it might be seven or fourteen days.' The need to wait until concrete set properly often caused some argument between the contractor's engineers, keen to proceed as quickly as possible, and the consulting engineers who wanted things to be according to the specification. The men mixing the concrete were always tempted to add more water as it made the mixture flow more easily but this carried the risk of later segregation and a weakening of the setting compound.

Plate 44.
The vertical tower in the centre comprises the cement elevator (the dark structure on the right) and the batching plant (the pale tower on the left) at the Cluanie dam site. On the left can be seen the cableway to convey skips of concrete to the dam. A quarry is in the background and the long, low shed houses a block-making plant (*Mitchell Report No. 1. Ronald Birse*).

Another factor that affects the strength of concrete is honeycombing. In theory, there should not be any but in practice it is sometimes difficult to prevent the grey mass from hardening around pockets of trapped air, resulting in a weak structure. It can also be impossible to spot until it is too late. Later in his career, while he was teaching civil engineering in Edinburgh

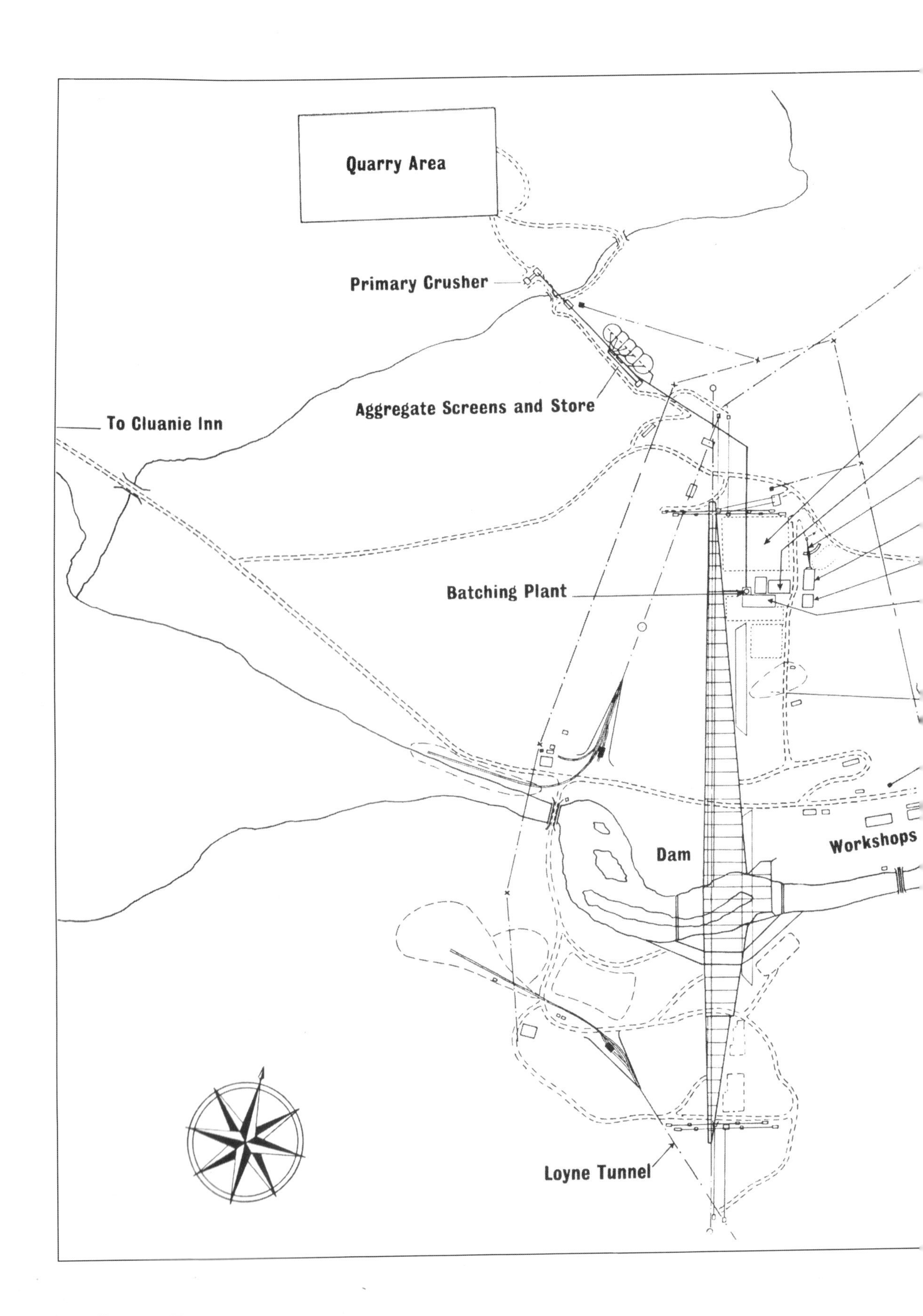
Quarry Area
Primary Crusher
Aggregate Screens and Store
To Cluanie Inn
Batching Plant
Dam
Workshops
Loyne Tunnel

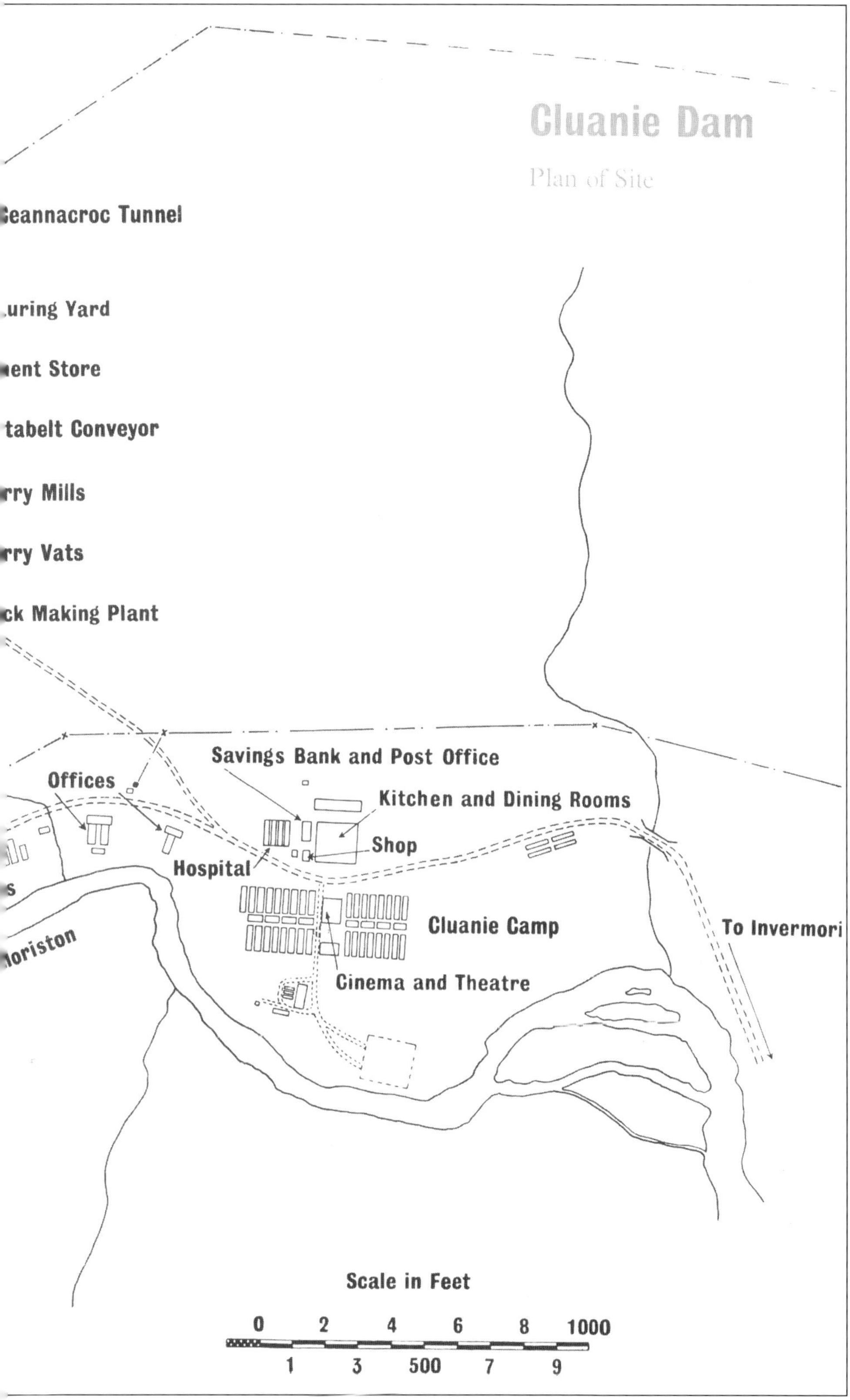

Cluanie Dam

Plan of Site

eannacroc Tunnel

uring Yard

ent Store

tabelt Conveyor

rry Mills

rry Vats

ck Making Plant

Savings Bank and Post Office

Offices

Kitchen and Dining Rooms

Shop

Hospital

Cluanie Camp

To Invermori

oriston

Cinema and Theatre

Scale in Feet

0 2 4 6 8 1000

1 3 500 7 9

Fig. 4.

A ground plan of the construction site and work camp for the Cluanie dam (*Mitchell Report No. 1*. *Ronald Birse*).

University, Ronald Birse was at times called on to investigate building failures – and sometimes the failure would be due to honeycombed concrete which nobody spotted at the time. The hydro scheme engineers had no means of detecting it, no ultrasonic equipment, and relied on experience.

'Honeycombing certainly was not welcome', said Roy Macintyre, 'and we always had to take steps to prevent it. By the time I was on the scene, it was better to show any to the consulting engineer and say can we clear this up. In the worst cases, you had to cut out a section and replace it, but these sections were not more than a quarter or half a square metre. I don't remember bigger areas on the dams being replaced, although there were bigger areas in tunnel linings, which were fairly thin. In a dam the main reason to replace honeycombed patches would be to prevent frost getting in which could disintegrate the concrete. Patching was an acceptable way of dealing with it.'

The ratio of aggregate – sand or gravel – to cement in the mixed concrete is also a crucial factor; the more cement, the stronger the result. Other problems arise from the fact that as concrete hardens it shrinks and generates heat. This rise in temperature in large masses of concrete can be high enough to crack it, and sets a limit on the amount of fresh concrete that can be laid at one time. To allow faster laying, American engineers developed a system whereby they could circulate iced water in pipes through newly laid concrete – the pipes would eventually be grouted up – but this technique was not used in the Highland dams.

Running water posed other headaches for the engineers. Water percolating through the rock lining a tunnel could be a considerable problem. 'It might have to be caught – channelled by corrugated iron – and then you would put in an arch of concrete, with the water still running above it, and finally drill holes in the lining and inject cement grout to fill the gap,' said Professor Birse. 'By and large, water being what it is, you couldn't stop it, you had to cope with it.' At least in the tunnels that simply channelled water over long distances, the load on the concrete could be low and, in such cases, usually there would be only concrete on the floor with bedrock sufficing elsewhere.

'As consulting engineers, we all had our own little patch. As I had developed an interest in the properties and behaviour of concrete, the contractor wanted to test the use of fly ash, the power station residue from burning coal. It's also called pulverised fuel ash, or PFA. It was described as potsolanic in its properties, a term that comes from the district of Potsola in Italy where volcanic ash is a natural cement. In fact, it was used by the Romans in their concrete.

'PFA was a waste product of power generation – produced when pulverised coal is burned in steam boilers – and the contractors decided it would be more economic for them to use a proportion of it in their concrete, particularly in the parts of the dam where high strength was not needed and where there were no severe stresses. There was a positive engineering advantage to using it as well as an economic one. The fly ash arrived sintered together to some extent and we ground it on site to incorporate into the concrete. We needed lab tests to find out how it would behave.'

The addition of fly ash made the concrete less permeable to water, although it took longer to set, and it was used in the building of the Lednock, Lubreoch and Giorra dams. Lednock represented the first major use of fly ash in a construction project in Britain. Lednock also has diamond-headed buttresses – the Errochty is the only other Board dam with them – to provide extra strength to cope with tremors along the Highland Boundary Fault that ran almost slap through Comrie five miles to the south.

The Board was always looking for innovations in dam design and construction that could cut the high costs. Mitchell Engineering Ltd introduced the Trief cement process in the building of the Cluanie dam in Glenmoriston in 1953, and also made use of pre-cast blocks on two dam faces. The Trief process, named after its Belgian inventor, Victor Trief, used waste slag from blast furnaces to replace in part the amount of Portland cement in its concrete mix. Lorries brought loads of slag from Colville's steelworks in Glasgow to the dam site where it was ground into a wet, creamy slurry at the rate of 170 tons a day. This slurry was then pumped to the batching plant to be mixed with sand, aggregate and cement to produce the final concrete. There were many advantages to using Trief cement; it can withstand the cycle of frost and thaw, is more resistant to water erosion and shrinks less as it sets.

The need for economy dictated that the Loch Quoich dam be made by the rockfill method. This technique, in which the dam is built up in layers of locally quarried compacted rubble around a concrete core, needs less cement than the traditional mass gravity design, and very little shuttering. When it was completed, the Quoich dam became the largest one of its kind in Britain and its rubble downstream face reminded one observer of 'a gigantic drystone dyke'.[30] The upstream face was covered in twenty-feet square concrete slabs, between twelve and fifteen inches thick, to create a waterproof skin. The rockfill structure ensured a massiveness not found on more conventional dams: the rubble fill forms in cross-section an immense triangle, 290 feet thick at the base, narrowing to a roadway at the top. 'The scale of the Quoich operation was incredible back in the 1950s,' said Laurie Donald who worked there in his first job after graduating from Aberdeen

Plate 45.
Working in the penstock access manhole on the Cassley section of the Loch Shin scheme, April 1959 (*NOSHEB*).

University. 'It was an American-type operation, especially with plant and equipment, very much in advance of what they had in this country before. Huge machines. The rock was laid in layers according to size and vibrated to pack it down – the ground shook.' The Vaich dam in the Conon Valley scheme was built with earthfill around a concrete core.

One of the more unusual professions associated with the schemes was that of photographer but on all the construction projects official photographers took series of pictures to record progress. Stanley Mills was one of them and he still recalls humping plate cameras over steep hillsides in all weathers to take photographs from the set vantage points he used to chart the progress of the work. Once his assistant disappeared from sight when he dropped into an unexpected bog hole. On another occasion, at the holing-through of the Clunie tunnel, Stanley crouched behind a row of sandbags with an explosives man as blast and debris roared towards them. Some of the pictures were touched up before inclusion in official Board publications: for example,

Plate 46.
Building the Duchally generating station on the Loch Shin scheme, November 1957 (*NOSHEB*).

builders' huts and scaffolding were painted out and a neat wall and gate inserted in the photograph of the sweeping panorama of Loch Lomond that graced the brochure for the opening of Loch Sloy. Another early picture of Loch Faskally had to be doctored to show a natural, tree-lined shore where only tree stumps and mud existed at the moment the photograph was taken.

The construction of the most northerly of the hydro projects, the Loch Shin scheme in Sutherland, was officially opened in a ceremony at Lairg on Tuesday 6 July 1954. In keeping with Tom Johnston's exercises in public relations, buses were laid on to bring spectators from Dornoch and Golspie, and Jessie Murray, representing the fifth generation of her family to have a croft at Drumnahaving, was accorded the honour of firing the first symbolic explosive charge. A crowd of over 1,500 gathered in the bright sunshine to listen to the speeches and the pipe band from Bonar Bridge. The press and television cameras were also in attendance.

A Board engineer was in charge of the catering and some of his men were

Plate 47.
The Creanich access road and camp on the Loch Shin scheme, March 1956 (*NOSHEB*).

Plate 48.
The Loch Shin scheme: the Glen Cassley access road, July 1955 (*NOSHEB*).

Plate 49.
Opposite. The Lairg dam under construction, March 1957 (*NOSHEB*).

Plate 50.
The Lairg dam under construction, March 1957 (*NOSHEB*).

pressed into service as stewards, handing out drinks and sandwiches. Ben Bentley and some of his pals from Dingwall had come along to set up and operate the sound system but were told on no account to touch the drink until the guests had been satisfied – 'If you do, your feet won't touch the ground.' 'One of the hydro boys worked out that, once the official business was underway, he could get "a cup of tea" without milk,' said Ben. 'He was carrying out these trays of cups, supposedly tea without milk, under their noses. The boss saw the trays with tea but thought nothing was amiss. By the time the thing ended, it was really like Fred Karno's army. An AA [Automobile Association] man, who was never known to imbibe, this day went daft. We had to get him and his AA Landrover home, and we did it – luckily he didn't have a call out on the way. His wife said "What happened to him?" and I said, "Oh he just had too much tea without milk".'

In his speech at the opening, Tom Johnston regretted that the Duchess of Westminster, the major landowner in the area,[31] had been unable to join the celebrations, adding: 'It would indeed have been significant of a new deal and an augury of brighter days in the north if under the aegis of the Hydro-Electric Board we could have had a woman crofter and duchess landowner on the same platform, cooperating in the same great cause.'[32] Johnston referred to the 'pioneering achievements' of the Westminster estate and expressed his belief that electricity would bring about a 'drift north' of industry and workers. In the meantime the crowd could look forward to the scheme needing 800 workers and the assurance that, for a time at least, there would be the prospect of full employment. In the evening, Miss Murray was invited to the home of a Golspie man to see herself on television.

[3]

'... Bonus, Bonus ...'

The tunnellers – the tunnel tigers – were the elite of the workforce and they knew it. If a man was a tiger he was worth something. 'It was a great honour to have that name,' said Hugh McCorriston 'It was a great thing to get into the tunnels. The tigers were paid more and we were on bonuses as well.' In 1955, on the Breadalbane schemes, they enjoyed a basic wage of £15 a week but this could be doubled by overtime.[1]

Once the exact position of the tunnel mouth had been fixed by the engineers, the tigers wasted little time in starting to dig. After they had scraped away the soil and the loose gravel to start one of the Affric tunnel entrances, the drillers found a huge boulder blocking their path. This was not seen to be a problem. The drillers noticed it had a flattish top, climbed up and proceeded to drill two concentric cones of holes. Explosive charges were packed in. The idea was to fire the inner ring with the outer ring going off a fraction later; it was accepted that this pattern would shatter the boulder. Unfortunately someone mis-wired the explosives, the outer ring went off first and blasted the core of the boulder skyward. It had gone some thirty feet up in the air before the inner ring of charges fired. The story goes that not a single vehicle or window in the vicinity escaped the fusillade of stony shrapnel, but no person was hurt.

Hugh McCorriston arrived on the Fasnakyle scheme in Glen Affric with his father, when the latter, an experienced explosives expert, a powder monkey, was called to come over to join the work team. The young Ulsterman became a member of the tunnellers, working twelve-hour shifts alongside the drillers, the muckers out, the loco drivers and the others, thirty to forty men on a shift at one time, from 8 in the morning until 8 at night. At times, the work did not end there. 'I've seen me coming out the tunnel after the Saturday nightshift, and they had a batching plant outside, mixing the cement, and half the boys hadn't turned up because they'd been in Inverness the night before. "Will you come back up and work today on the outside?" Well, that wasn't too bad. We'd go down to the camp, get our breakfast and a packed lunch, and come back up, work twelve hours outside,

Plate 51.
A Ruston diesel locomotive, one of fifteen that carted spoil from the tunnels on the Breadalbane scheme (Civil Engineering and Public Works Review, *No. 594, Dec 1955. Ronald Birse*).

Plate 52.
A Holman SL.200 drill in action at the break-in of the tunnel that will convey water from the upper reaches of the Beich Burn towards the St Fillans generating station on the north shore of Loch Earn (Civil Engineering and Public Works Review, *No. 594, Dec 1955. Ronald Birse*).

go down, get our dinner, come back, and work the Sunday night as well – thirty-six hours straight through. I was fit then. We did that quite often.'

The tunnel reached into the slabs of rock like a gateway into the netherworld. Electric bulbs were strung along the roof, and railway tracks for the bogeys that carried the spoil ran up the floor, through the inevitable pools of dirty water, mud and rubble. Pumps tried to maintain a movement of fresh air into the enclosed space, reeking of gelignite and diesel exhaust, but this was at times a pretty Heath-Robinson kind of affair, according to Don Smith, who worked in the tunnels, maintaining all the electrical equipment including the compressors and transformers, and extending the lines of light bulbs as the drillers dug deeper and deeper into the rock. Sandy Payne remembers the tunnels as being the nearest thing one might imagine to be Hell – 'noise, lights, dust, the black-faced people appearing and disappearing in the gloom'.

'Very often we were called into the tunnel and would spend maybe a couple of shifts in there,' recalled Archie Chisholm. 'You were in darkness all the time with a dim light. I'll always remember one weekend, one Saturday, Davie Main and myself went in to repair this hoist. We had to put new bearings in it. A double shift. We came out on the Sunday afternoon. It was a bright day of summer, and it was like coming from Hell into Paradise. That was the feeling it gave you. A Technicolor picture after being stuck in the tunnel.'

'The lights? If the bulb was more than six feet away you couldn't see it,' recalled Iain MacRae of the Glenmoriston tunnel. 'It was drastic – you couldn't see your nose in front of your face for diesel fumes. The tunnel went

Plate 53.
Inside a tunnel on the Clunie scheme. Rail lines and cables snake along the rubble-strewn floor under the glare of the electric lights (*Mitchell Report No. 1. Ronald Birse*).

in for four and a half miles and they had an adit midway between the dam and the outflow. There were fans but they weren't very good. We went in only periodically when there was a bit of shuttering to do but for those who worked there all the time I don't know how they survived. Underfoot was not too bad – near the surface the water was fairly widespread but the deeper you went the drier it became. But the stoor!'

Iain Macmaster's duties included extending the cables of lights and once he had to do it as the shift was drilling. 'It was a noisy place. They didn't wear ear plugs, or only used them when a shot was fired, so they were carrying on. I was in a strange environment, in this terrific noise, and rocks were falling. I says there's only one way to deal with this and I might as well do it now and I went back fifty yards, put the pliers through the wires and left the shift in darkness – they had to stop. I think they knew what I had done but they didn't bother me again. After that when I said stop until I got a wee bit finished they would stop.

'If you're working on your own, there's a very "loud" silence in a tunnel. You might hear a splash – a stone falling from the roof into a pool of water – and you think, I wonder if there's going to be a fall today and who will dig me out. I was only twice on the wrong side of a fall – I couldn't take the loco out because the line was blocked but I could walk out. You say to yourself, if that much can fall, how much more can fall. There was an eerieness when you were on your own. It was amazing how varied the rock was. They used to ring the tunnels in steel to prevent falls but all in all I suppose most places had good firm rock, pretty stable.'

Plate 54.
Wodek Majewski and colleagues in his tunnelling squad (*Wodek Majewski*).

Wodek Majewski began to learn the tunneller's job at Mullardoch on the morning after he arrived with his fellow Poles. 'They gave us helmets, boots and oilskins, and one of the foremen took us to the tunnel. There were four of us in the one squad. Mates used to stick together, you know. The tunnel was quite big and there were these big, heavy machines, Drifters they called them, that needed two men to lift, for drilling, and the noise. We were going one step forwards, three steps back. After the first couple of hours there was a teabreak outside and everybody was saying "A couple of weeks, I'll make some money and be going back to England", but we just carried on and we lasted a year and a half. Then, we got our first pay and of course the pay was quite good at that time.'

The American-made Drifters were operated by two men – the machine man who controlled the drill and his mate, the spanner man, who changed the tungsten drill bits as required. The Drifter weighed about fifty kilos and was powered by compressed air. As it hammered at the rock, water was pumped in through a hose to wash away the loosened debris. The spanner for changing the bits weighed about eight kilos. 'The screw-in bits were

Plate 55.
Tunnel tigers in Strathfarrar having a break, 9 December 1959 (*Press and Journal*).

changed every two or three feet as you were drilling, or perhaps oftener, depending on the rock,' explained Otton Stainke who, like Wodek Majewski, had not set hands on such a machine until he joined the team in the Mullardoch tunnel. 'To drill two or three feet would take two or three minutes. Once a start had been made on the hole, you changed to a four-foot drill and so on. The point of the bit was shaped like a star and it could be resharpened until it got too small. The hole had to be about one and a half inches in diameter to take the charge. We drilled to eight or ten feet before the charge was laid, or if it was good rock perhaps twelve feet.'

The noise made speech impossible and the machine man and the spanner man communicated through hand signals, an open hand meaning slacken off and a clenched fist tighten up. In the early days when the Poles were still learning, Wodek Majewski remembered a mate mistaking the clenched fist for a threat and offering to fight the machine man. Another spannerman, from Newfoundland, at first thought the machine man's open hand meant he wanted to shake hands with him. 'You soon got the feeling of the machine', said Otton Stainke, 'and learned to control it. It was fixed on the gantry, and

Plate 56.
A conveyor belt carrying concrete in the Clunie tunnel, February 1949 (*NOSHEB*).

Plate 57.
Opposite. Horseshoe travelling shutter in the Clunie tunnel, December 1948 (*NOSHEB*).

underneath there was a twist grip [by which you could] put more pressure on it or less. Too much pressure would make it stop, and the drill could stick. You kept pressure to keep the drill turning. Every man had his own machine; if he had a good one, he tried to keep it, and if somebody else went off with it there would be a row. The same machine men and spannermen always worked together.'

'In drilling everybody was in a hurry – bonus, bonus,' said Wodek Majewski. 'By right you should be starting with a three-foot drill and then change to eight feet, using the three-foot hole to guide the longer bit. But at the end we started with the eight foot, with one man holding the end of the drill while you go slowly. It's turning in his hand but there's no need to hold it hard. When you have a hole about an inch deep he can let go and you can turn the water on and carry on. It took up to half an hour to drill the full eight-foot hole, depending on the rock, some was hard but some was kind o' soft. Usually the whole face was ready for blasting in about one and a half to two hours. Drill bits broke a few times. I nearly killed one chap. The bit just snapped. Too much pressure. The man was holding the drill and when it snapped I and the machine just fell forward. I avoided him and it was all right, no harm done. There were a few accidents happened like that, you know. We used to get about £10 or £12 for a shift. The time we broke the record we got £55 bonus plus the wages.' If a drill bit stuck it would have to be cut, and the machine man would have to start again making a new hole

Plate 58.
The plug under Loch Fannich wired for blasting, September 1950 (*NOSHEB*).

Plate 59.
Opposite. A skip of concrete being poured behind the climbing shutter in the lining of the Fannich surge shaft, May 1950 (*NOSHEB*).

beside the blocked one. Improved drilling machines, such as the Atlas series, became available on later schemes. They were lighter and were operated by one man.

The drillers worked at the tunnel face, standing on two or three levels of moveable scaffolding called a gantry or jumbo. When the drilling was done, the gantry was run back on rails for about 500 yards. '[The wall] was all charged up – that was what my dad was doing, packing the explosive into each hole. Then we pressed the plunger, and that brought the whole lot out,' explained Hugh McCorriston. 'We didn't go very far down the tunnel [away from the blast], it took too long to go back. We were supposed to stay out an hour before we went back in, but that never happened, there were no unions or anything to stop us.' Otton Stainke also learned the handling of explosives and the placing of the charges at Cannich.

The holes to take the charges were placed carefully. In the centre of the face was the burn cut, perhaps nine holes; another series of holes ringed the tunnel circumference; and a further row, called the lifters, was drilled along the foot of the face. When the face was fired, the burn cut went off first to crack the rock, followed a fraction later by the circumference charges, and finally the bottom lifters exploded to lift the wall of fractured rock and drop the whole loosened mass. 'They had a galvanometer to check the firing circuit was good,' said Don Smith. 'If they didn't get continuity – sometimes there was a bad detonator, and it wouldn't give a reading – then they had to check it. You had to have three or four detonators wired in series, otherwise the 1.5 volt charge would put a single detonator off. You had to be very careful in checking, if anything was wrong. You were taking wires from here and there to make sure you had enough in line.'

The gelignite used in blasting was not pleasant to handle. The long, thick sticks, each about half a pound in weight, came well wrapped in greaseproof paper, but they stuck to the hands, gave off fumes that 'could cut the eyes out of you' and were prone to sweating, which, if it happened too much, could be dangerous.

When the holes were drilled, the sticks were packed in tightly, a detonator was added to the last one and the whole assembly was stemmed with sand, until the wall was primed, wired and checked, ready for firing. 'It never failed to go off, we were quite lucky that way,' said Hugh McCorriston. His father had once been badly injured in Ireland, by the proverbial damp squib. It was difficult to count whether or not all the laid charges had gone off and, on that occasion, one fuse had been too slow with the result that Mr McCorriston Senior had been walking back into the firing zone when the delayed explosion occurred. He had been knocked unconscious for eleven days.

Plate 60.
A change of shift on the lining of the Fannich surge shaft, May 1950 (*NOSHEB*).

'Misfires weren't frequent, but I'll tell you what was bad,' said Don Smith. 'When there was lightning in the area. The lightning could hit a rail, travel up the tunnel to the face and blow it up. That didn't happen here but it had happened in other places, so the rule was don't charge up the face while there's lightning about.' When a charge failed to fire, a blowpipe was used to blow out the sand to expose the faulty explosive.

Archie Chisholm remembers 'the sweet, pungent smell' after the gelignite had been fired, 'a bit like cordite but more heavy, and it sticks to your clothes. It was the predominant smell in the hut at night when the shift came off. You could smell the tunnel in the huts. Sweaty feet and gelignite smoke. If you were in [when they fired] you would feel the blast of air coming right through, with the smell of geli and diesel fumes. If you took out a match, you

28/52 Glascarnoch : Luichart : Torrachilty Project 8/8/56
Glascarnoch Tunnel
Start of walls and arch lining using 200 ft.
telescopic steel shutter

Plate 61.
Opposite. Glascarnoch tunnel: start of walls and arch lining using a 200-foot telescopic steel shutter, August 1956 (*NOSHEB*).

Plate 62.
A telescopic shutter for lining a tunnel with concrete, in the Killin Lochay Tunnel (*Edmund Nuttall Ltd*).

could hardly see which was the right end. Grey dust came off the rock and the men would be spitting out grey spit, just like cement. The worst thing to do was wash in warm water because this made it stick. A cold shower was the way to get it off initially, otherwise it would go into the skin and become claggy. A cold shower and then a hot shower. A lot of the boys of course just came off shift, had their meal and lay on the bed.'

After most of the rubble and dust had settled in the wake of the explosions, some of the gang began to hit the tunnel roof to knock all the loosened rock down. Then the EIMCOs, small machines driven on rails by compressed air, were used to clear away the rubble and debris after a blast – some of those drivers 'could make the thing talk', remembered Don Smith with admiration. The EIMCO, controlled by the driver standing on a side platform, had a shovel on the front for scraping up the spoil and tumbling it into a row of skips or bogeys on the railway track, along which an electric loco pulled the full skips to the tipping bing in the outside world. Paddy Boyle drove an EIMCO: 'You had a mate who held the compressor bag so you wouldn't run over it on the rails. It had hand controls – one lever to go

Plate 63.
Constructing the access shaft to the Glascarnoch tunnel intake, September 1956 (*NOSHEB*).

Plate 64.
Two men in the Clunie tunnel enjoying a cup of tea (*Perth Museum. Copyright: Louis Flood*).

Plate 65.
Break for refreshment in the tunnel (*Perth Museum. Copyright: Louis Flood*).

Plate 66.
A holing through ceremony, probably taken in the Clunie tunnel (*Perth Museum. Copyright: Louis Flood*).

Plate 67.
Opposite. Excavating the sump under Loch Fannich, July 1950 (*NOSHEB*).

forward and back, and one to empty the bucket over the back of you into the skips.' When the mechanical shovels had done all they could, the men cleared away the remainder with hand shovels.

'The Cannich boys broke the record for Great Britain and Europe for the amount of rock taken out in one week,' said Hugh McCorriston. 'They were on a bonus system, going hell for leather, no hanging about. The whinstone is tough, tough stuff.' The week's record is recorded for a drive of 126 feet. In all, over three million tons of rock were excavated to drill the fifteen-foot tunnel through the three-and-a-quarter miles between Mullardoch and Benevean, and 253 tons of gelignite were fired off.[2]

The man in charge of the work was the tunnel boss. 'The tunnel boss was really your man on site. If you fell out with the tunnel boss, he could make your life difficult. Some were low key, others were characters indeed – well, they were all characters but some were that in a quiet, laid-back way,' said Iain Macmaster. 'The majority were Irish. The first one I met was Irish – at Lochay, and he had gravitated there from Sloy where he had survived an explosion. He carried so much clout the site engineer wouldn't cross him. Very often the tunnel boss's decision was the one adhered to – if a tunnel boss had the experience any site agent would think long and hard before disagreeing with him. There was a famous Polish boss on one scheme. I liked him but he had a very, very tight regime. A few people had problems with him, and went down the road. There was a rumour that he had been in the SS and, knowing the guy, I wouldn't have been surprised.'

Sir Duncan Michael remembers a boss who came from Avoch on the Black Isle: 'He had the day shift – a John Wayne type, a big, calm man, never shouted, just worked, and he had more Scots in his team, and they always seemed to do well. Maybe they left the crap for the night shift – we'll never know. After all you can make the progress and leave all the pump cleaning and so on. Now the night boss was a thin, ginger guy with a swagger – a Clint Eastwood rather than a John Wayne – and he always had a ragbag of a team somehow. I knew him and always gave him the best shot when I could. But the teams were different.'

Keeping the tunnel on the right course underground was the responsibility of the engineers. Now it is done with lasers but on all the major hydro-electric schemes the engineers stuck to what was called the theodolite and candlepot system. Accurate observations from the original baseline determined the position of entry to the tunnel and the angle at which the drillers had to excavate. At regular intervals, usually every 300 feet, a steel plate was embedded in the floor with the precise tunnel centre line engraved on it, and above each of these floor markers a wooden strap was fixed to the

Plate 68.
EIMCO loaders shifting rubble in the Cruachan access tunnel (*NOSHEB*).

tunnel roof. A chain was hung from the centre of the strap and at its end burned a candle in a metal tube. By aligning at least two candles and a third at the centre of the face, the day-to-day drilling of the tunnel could be kept on course both horizontally and vertically by the crew. 'It was a good old reliable technique,' said Bob Sim. 'The candle burned with a point to its flame and that, with frequent measurement by theodolite, allowed an accurate measure of the angle we had to keep to.' The rate of progress in excavating a tunnel was such that a new candlepot marker had to be put in place more than once a week.

Boring through bedrock required hardly any shoring up of the exposed walls, a feature that risked the occasional rock fall but generally allowed fast progress. Tunnels had, however, to be scaled, trimmed to the right size, after the engineers measured the diameter and found that it was not quite big enough to accommodate, say, a three-foot concrete lining. The trimming was done by hand tools. 'That was when it got a wee bit dangerous because of loose rock,' said Hugh McCorriston. 'A piece dropped on me but it didn't do any damage. We had no helmets. When I went there first, you were lucky if they gave you a pair of Wellingtons. You had to bring your own oilskins and your own boots, and then one or two got a wee tap on the head and eventually they did issue us with helmets. Bodies were plentiful, there was no shortage of anybody wanting to work; they may not have stayed long but they came there.

'The tunnel could feel warmer than it was outside. The heat in the earth. And you were working so hard. Eventually, because there was a lot of water dripping in, you were mostly in oilskin jackets and trousers, and that's pretty hot to work in. We were quite delighted when they did break through the tunnel and the air started coming, circulating through. There was no fog then, you could see through it. If you were in there on a foggy night, boy, you've no idea – it was thick in the tunnel.

'The winter in the tunnels didn't seem to bother you too much. I don't know, you're young and fit, and I don't think I ever had a cold or flu. And you can soon get a sweat up in the tunnels. We were well over a mile inside. Eventually, once the tunnel was right through and the shaft was down, you were getting air circulating right through. Until the tunnel was holed through, we walked back and fore, but then they put in bogey lines. We had these little trolleys like jaunting cars – three or four men sat on either side – and I think that was more dangerous than actually working in the tunnel, the speed these boys used to take you in and out at. They were diesel-operated locomotives. Mostly they were driven by the Poles. The Poles preferred working with machinery, they didn't want the pick and shovel. When they were coming out, they'd be in a hurry to catch the lorries back to the camp and they would just throw the loco out of gear, and freewheel out. It was very, very steep. The whole tunnel from one end to the other, about two and a half miles, had a drop of 500 feet.'

Occasionally the work at the face was interrupted for a diverting reason. Once, in the approach to a surge shaft, the tigers uncovered two petrified trees lying at an angle in the rock wall, and work stopped for a fortnight until palaeontologists came to examine the relics of life from millions of years before. Official visitors and photographers were also likely to drop by. As an electrician, Don Smith was looking forward to wiring up lights once for photographers at Cannich but was fascinated to find that his services were not required: the cameramen set up polished aluminium reflecting panels in the mouth of the tunnel and guided brilliant natural light into the hill for three-quarters of a mile to the workface.

Different techniques were used to drill some of the surge shafts. Ronald Birse described one: 'We had this interesting system which had been developed in Sweden, I think, for excavating the vertical shaft. If you do it from the top, working down and down, it's a very tedious business to excavate a shaft about 500 feet deep. The Swedes developed a system for excavating upwards. They started by drilling down a six-inch-wide hole. That took some doing, to keep it plumb, and they then put a winch at the top, fed a steel cable down, and cut upwards from a platform at the bottom. The rock spoil fell down under gravity and was mucked out from below. The upward drilling was more difficult – it needed more muscle – and more dangerous. Once the men drilled up six feet or whatever, they charged up the holes, lowered the work cage, took it fifty feet back in the tunnel, vamoosed, and let off the charges. I think there was one accident when they were doing that.'

A more sophisticated upward tunnelling device, the Alimak, was to appear from Sweden some years later, when it was used on the Ben Cruachan scheme.

The concreting of tunnel walls was done with a moving shutter and a Bugee pump. This large device moved on rails and worked by pumping liquid concrete by compressed air into the space between the shuttering and the exposed rock. Fish oil, or mould oil, as it was sometimes called, was painted on to the inner face of the shuttering to prevent the concrete from sticking to it, and added a further distinctive element to the mix of smells in the enclosed space. 'The pumping of concrete never stopped,' said Archie Chisholm. 'One shift took over from the previous one. There was a real panic station if there was a breakdown in a pump and the concrete stopped moving, as it would set in the pipes. A big bing of old pipes at Cannich yard were solid with concrete. The pipeline, a six-inch steel pipe, ran all the way from the batching plant to the face. The pipe sleeves had buckles on the end to join them up. If it choked, it could be unfastened quickly and the concrete blown out. The Bugee pump had a big funnel where the concrete was poured in, and big rams that you could hear miles away when the pump was going. Little shuttle valves took in air to work the piston to push the concrete through. It was pressed against the rough tunnel wall, where it would stick and set. All the Benevean-Fasnakyle tunnel was concreted except for the last bit – it was lined with steel. The Mullardoch-Benevean tunnel wasn't concreted.'

Milestones in the tunnelling were marked by celebrations. The teams of tunnellers competed to accomplish record drives and in May 1957 the *Perthshire Advertiser* carried an unusually detailed report on the claiming of the record session in the Lednock tunnel in the Breachlaich section of the Breadalbane scheme. The teams were working for the contractor R. J. McLeod and completed 1,621 feet of eight-foot-diameter tunnel in thirty days. The two crews were made up of the following Scottish, Polish and Irish personnel:

> Day shift – shift boss, five machine men, one loco driver, one EIMCO driver, two tipmen, three platelayers, one powder monkey, one compressor man/drill doctor, one fitter, one welder/ blacksmith, one electrician, and one handyman
>
> Night shift – shift boss, six machine men, one loco driver, one EIMCO driver, two tipmen, one compressor man, one handyman

Keeping the pace up around the clock, they managed four 'rounds' per shift. Each three-hour round comprised forty-five minutes of drilling, twenty minutes for charging and blasting the face, forty minutes for a meal break while the dust and fumes cleared, and seventy-five minutes for mucking out. The quickest round they managed, however, was only two hours fifteen minutes; one imagines an anxious shift boss with one eye on his watch and other on the men, urging them to go for it. At this rate, the best advance in a week was 429 feet.[3]

Hardly had this achievement been published than it was being challenged. In June 1956, though they hadn't bothered to establish a claim to a record, Mitchell Ltd's men in the St Fillans section had driven 1,837 feet in thirty days; and in October 1955 another team working for Mitchell Ltd had

Plate 69.
The record-breaking team who drilled the Loch Sloy tunnel (*courtesy Edmund Nuttall Ltd*).

carved out 557 feet in seven days. In Glenmoriston the average rate of progress was 160 feet per week, with a maximum of 230 feet being achieved once.[4] The three-mile tunnel from Cluanie to Ceannocroc was dug in thirteen months.[5]

Holing through, when the teams of tigers drilling towards each other finally removed the last rock wall between them, was a time to break out the beer and possibly pause for a group photograph. The Clunie tunnel was holed through on 30 August 1949.

'Three hundred and fifty feet below Cammoch Hill near Pitlochry a page of progress was written into history yesterday with 600 lbs of gelignite and a piece of tartan ribbon,' reported the *Perthshire Advertiser*. When the deputy agent of Cementation Company Ltd fired the last shot to blow the final nine-foot barrier to smithereens, it had taken exactly three years for the 300 workers to bore the tunnel from each end. The length was a little short of two miles and the two sections were less than one inch out when they met. After the dust settled, a tape of Royal Stuart tartan was hung across the gap and ceremonially cut; beer was brought in; and the foremen of the two teams – Harvey Jessiman from Aberdeen, and George May who was shortly to return to Toronto to drill subways – shook hands.[6]

'It was amazing how quickly the tunnels were completed,' said Archie Chisholm. 'For example, they started the six miles of tunnel between

Mullardoch and Benevean, and between Benevean and Fasnakyle in 1946 and it was done by 1949. They had broken through, drilling from both ends. There was only about three feet of a gradient in the three miles between Mullardoch and Benevean, just enough to make the water flow, and, when they broke through, they were only inches out.'

Sometimes a guest was invited to perform the completing blast, such as in the main Fannich tunnel on 3 March 1950 when the wife of J. Guthrie Brown, a partner in the consulting engineering firm of Sir Alexander Gibbs, fired the last shot. The destruction of the last rock wall about one and three-quarter miles inside the mountain was heard as a dull rumble. The two tunnels met with an error of less than one and a half inches. Sir Hugh Mackenzie, for the Hydro Board, cut a tartan ribbon across the adit entrance and led a party of visitors through the tunnel, through the newly blasted and tidied hole and on for another mile until they emerged south of Loch Fannich.[7]

Five months later, the Fannich scheme was the scene for a more spectacular holing through. The tunnel to carry water from Loch Fannich to the Grudie Bridge power station was excavated three and a half miles up the gradient until all that remained of the bedrock was a plug between the end of the tunnel and the bed of the loch. The last charge was designed to blow away the plug and allow the loch water to flow down the tunnel. Shortly after midday on Thursday 7 September, Lady MacColl pressed the button to complete the firing circuit.

Just before this, her husband had addressed the assembled guests. He predicted they would see a water spout rise up in the loch about 100 yards out and added that he shuddered to think what might happen if this did not materialise, that some of them might land in Loch Maree. The *Courier* correspondent went on: '[The party] heard two muffled explosions and watched the loch's surface, already ruffled by a strong westerly wind and driving rain, lift at one spot to a height of almost sixty feet. When the water spout receded the only sign which remained of the explosion was a gravel-coloured patch of water upon which floated a number of dead brown trout which, however, were soon picked up by marauding gulls.'

The event, baptised by the junior engineers as 'operation bathtub', had demanded careful planning. The plug left at the end of the tunnel was fifteen feet thick and a special sump had been excavated so that, when the 400 tons or so of rock were blown to smithereens, the fragments would fall out of the way. Steel and concrete bulkheads had been placed in the tunnel 500 feet down from the intake to contain the inward rush of Loch Fannich water. When the scheme was completed and the turbines at Grudie Bridge were ready, the bulkheads were removed to allow the water to flow on its full journey.

The Glascarnoch tunnel was holed through on 30 October 1954, a successful end to two years of drilling by A and M Carmichael. The five-mile tunnel was the longest to be finished without adits at that time and had required the shifting of 183,000 tons of rock and the firing of almost half a million pounds of explosive. Sir Hugh Mackenzie fired the last shot from a switch on a table draped in a Mackenzie tartan rug and set up at the south tunnel entrance, five miles from Garve on a steep hillside overlooking the head of Loch Luichart. The explosion, two and a half miles inside the hills, was heard nine seconds after the switch had been depressed. It must have been a long, anxious moment for some of the engineers laughed when one of them said 'I'm certainly glad to hear that'. Wearing helmets and oilskins, the party walked into the tunnel and took a train to the explosion site. As they rode for twenty minutes to reach the spot, the ventilating pumps laboured to extract the dust and fumes and by the time they arrived the air was clear: it was 'a somewhat eerie experience' recorded one observer. The north and south drives had met exactly.[8]

[4] 'You wouldn't call the King your uncle'

On a cold October evening in 1946, Patrick McBride saw for the first time the site for the Clunie dam, ten miles up from Pitlochry: 'We were introduced there to three buses on the side of the road, two for sleeping in and one for the cookhouse. We had no running water, we had no light, and very little transport in them days, just one small van. The buses weren't lined inside, they were just ordinary buses with three bunk beds on each side and a little stove in the middle. If that fire went out during the night, you jumped to get some sticks on it. Of course our first job there was to build the camp, an office, and a store. There were twelve of us on it, and two of us were the cook and the fellow who did nothing but carry spring water in a bucket to keep the cookhouse tanks full. We had a Tilley lamp and it was like a gypsy camp, just sitting at the side of the road. There was a pub further up, the Tummel, and if we got a lift we went up there for a few pints at the weekends.'

The twelve men set to work to build the camp for the workers who would arrive to start on the scheme itself. With no machines to help them, they laboured to clear away undergrowth and level the ground. Lorries began to arrive daily with loads of wooden hut sections and with them came the snow, the 'terrible snow', recalled Patrick. 'If [the lorries] didn't make the site we had to go down the road and clear away the snow – we'd got a little tracked machine at this time. We cleaned the road, pulled these lorries out of the dyke, got them out, got them up to the camp. That snow lasted for months and months. We were out every morning and all these sections had to be carried to where they had to go.'

Unable to recruit more help because there was nowhere for them to stay, Patrick and his colleagues worked on. The project manager found living quarters in a hotel, and the works manager, Jimmy McBride, Patrick's uncle, had a small bungalow. Eventually, the men erected huts, a canteen and an office, and, after Christmas, electric light and pipes for the spring water were installed. The first workers, recruited in Buckie, Dundee and Aberdeen,

Plate 70.
The Fannich scheme: excavating a channel for a temporary diversion of the Grudie river, June 1948 (*NOSHEB*).

Plate 71.
Left. A Polish cook feeding a sheep at Mullardoch (*Don West*).

Plate 72.
Far left. An engineer at Mullardoch in waistdeep snow (*Don West*).

Plate 73.
The work camp at the Glascarnoch dam, February 1953 (*NOSHEB*).

29/7 Glascarnoch : Luichart : Torrachilty Project. 19/2/53
Glascarnoch Dam. View of site looking South-West.

Plate 74.
Constructing the Wester Aultguish work camp on the Glascarnoch scheme (*NOSHEB*).

arrived – at about half past five one bleak, cold evening. 'When they saw the place', recalled Patrick, 'they said "That's it, we're going". But they had to stay for the night, twenty of them. Some of them played poker and became skint, and in the morning they all left except five. Funny, but them five stayed with us right through the contract.'

Home for most of the workforce, at Clunie as on the other schemes, became the camp – a village of timber or corrugated-iron Nissen huts, thrown up on a level patch of ground like a frontier settlement in some wild gold rush. Most of the buildings were for sleeping quarters but beside them might stand the post office and savings bank, a shop, the kitchen and the wet and dry canteens, a theatre/cinema, a first-aid hut, and an array of offices and workshops. The accommodation for the women workers was nicknamed the hen-house at one scheme. A typical sleeping hut would have iron beds arranged dormitory-fashion on either side of the interior space, with perhaps a few cubicles at one end. Each man had his bed and a locker. Toilet arrangements, at least on the earlier schemes, could also be basic: a few Elsan chemical closets side by side with no partitions was one arrangement. At the work place even this modicum of comfort was foregone and men had to make do with a plank over a pit which, if nothing else, did much to enhance a sense of balance. It wasn't surprising that a handy clump of trees would commonly be pressed into service.

'As an eighteen-year-old in camp, I was just bewildered at first,' recalled Sandy Payne of his arrival in Glen Strathfarrar. 'I had hitchhiked down to Beauly and the keeper had given me a lift up and dumped me off. Fortunately a friend had got the job lined up, because men were queueing looking for work. I was given a bed number, blankets, sheets and a pillow case, and told to get on with it.'

'We made our own beds,' said Donald MacLeod. 'There was an orderly – he swept, but you were responsible for your own bed. The sheets were changed – I can't remember, once a fortnight. You went to the camp boss's office and you got a new sheet. Once a fortnight. Cold sheets, white. No flannels, nothing soft.'

At Cannich, the men were issued with one clean sheet every week. The old bottom sheet was taken off for the laundry, the old top one put underneath, and the new one became the new top sheet. The camp at Cannich stood where the village is now. Hugh McCorriston remembers it as being all Nissen huts, like a military camp. Only the women who made up the catering staff had wooden huts. The Nissen huts were heated by steam pipes that ran around the curving wall but it was not always very effective, especially if your bed was at the end far from the boiler. It didn't matter all that much. The men coming off their shift were usually so tired that they lay down to sleep and that was it. If it was cold you could throw an extra jacket on the top of the bed.

The cubicles offered a little more privacy. They gave a man his own space, or at least a space shared only with one other fellow, a bit more room and even a wardrobe for his clothes. There was a waiting list for cubicles but, if you stayed in a hut long enough, you could be assigned a bed in a coveted cubicle, even if the wall wasn't closed all the way up to the ceiling of the hut and the noise was just as loud. 'There was always some idiot marching up and down when you were sleeping, the big boots – bang – bang. It was rough but it was good. All you wanted was to eat and go to sleep, that was it,' said Hugh McCorriston.

William Rosie was amused by the behaviour of some of his companions:

'There was one bloke, he came in with a bonnet on, a scarf round his neck, boots on his feet, and he turned into the bed like that. I was amazed. I was opposite him. I said to myself he's never going to sleep like that, but he did. When he went to the washhouse in the morning he took the scarf off, gave himself a scutch roond his face, and that was him finished, ready for his breakfast.'

Eat and go to sleep – the priorities of most of the men arriving from their shift in mucky clothes, tired, ravenously hungry after maybe twelve hours of physical labour. Food became very important. And the wages. Friday was pay day when work stopped earlier than usual and everyone collected their reward for the week. 'We had to queue up for our wages at the pay office,' said Hugh McCorriston. 'They would shout your name, and they had a big tray with all the pay packets lined out. It was all notes and cash. I don't think you had to sign for it. Everybody got paid at the same time.'

This was the time to stock up on new clothes. A man might go through a pair of overalls in a couple of months and, unless they were issued with clothing, he had to replace them as best he could. Patrick McGinley remembers a Pakistani peddler coming round at weekends with a bulging suitcase, filled with shirts, shoes and underwear, and a whole lot more stowed in his van. 'He could talk Irish Gaelic to some boys, he was that used to it. A lot of the men came from the west, all Irish speakers. Nice bargains today, Paddy, he would say, and he charged the last penny for them up there in the camp. We had different names for him.' Many men put off replenishing their wardrobe until they had a weekend break in the nearest town.

'There was a post office and a wee shop in the camp,' said Patrick. 'I was good mates with a fellow, John McNulty, who worked in the shaft, and the two of us would go to the shop for bacon, a couple of pound, and a loaf of bread, and we would cook that ourselves for our supper. On the stove in the hut. You could fry up on it, maybe one, maybe two pans between the whole lot. If anybody wanted a lend of the pan you'd give it to them. Me and John used to go down to the shop. There used to be great bacon going at that time, in this wee shop in Invermoriston.

'We were getting a good wage at that time and there was nowhere to spend it. We were in the camp and that was it. I didn't drink much and didn't spend too much time in the canteens. The tunnel men – they used to drink a lot, they were heavy drinkers. Myself and another man by the name of James Bryson – I think he was a Beauly man – he was a great fitter – we got this owld motor bike, a BSA. We stripped it down to the frame and painted it black. He knew where you could get wheels and, between us, we got all the bits and pieces, all the parts. We fiddled away at it and got the BSA going. We used to go out for a spin to different places on it. It passed the time, it was something interesting instead of lying about the camps. He kept the bike when I moved on.'

One man on the Glen Affric scheme supplemented his pay handsomely by playing his accordion in pubs in Drumnadrochit. This private income allowed him to impress his hut mates one night when an idle discussion arose over the range of bank-note denominations. There was general disbelief when the accordion player announced there were such things as £50, £100 and even £1,000 notes until he opened his locker and brought out a shoebox full of unopened wage packets and the notes he had described. Remarkably, although his possessions were never locked away, no one tried to pilfer anything from him.

The comparative qualities of the camps – which was the best billet – remain in the memories of the itinerant workforce, moving from scheme to scheme through the late 1940s and the 1950s. A bad reputation became attached to the Calvine camp, where the huts were old and dilapidated – they had previously been used in the War by Canadian lumberjacks, and then to house German and Italian POWs – but most of the workers simply put up with the discomfort. In the drifting population old mates could suddenly find each other again. Many of the workers on the Glen Affric scheme had been labouring together at Lyness, the headquarters of the Scapa Flow naval base, during the War and met up again on this new venture. When he left Cannich, Archie Chisholm thought he had seen the last of the old Canadian blacksmith Elmer Turner, but this was not to be: 'When the Cluanie scheme started I moved to Mitchells and was issued with space in Hut 27. When I went in who was in the opposite bed but Elmer. We had a lot of escapades after that.

'I used to pick him up at his house in Balnain in Glen Urquhart on a Monday morning because I was living in Strath Glass then and had a wee Austin 7. We were up near Invermoriston one morning driving along, and this roe deer jumped off the bank on to the road, and hit us. Elmer cried "Stop the car". He always carried a big knife, and he cut the hindquarters off and put them in the boot. I didn't think much more about it. In the hut at night, quite late, I came in and fell asleep. A lot of the guys had a wee one-kW stove under the bed in case they wanted to cook up something. I didn't. About midnight, I got this nudge to wake up. Here was Elmer presenting me with a bit of this roe that he'd cooked in a pot under the bed. He used to sit on the bed in a red and black lumber jacket, wearing a hat, smoking a big pipe with a bull's head and horns, and he would sit back, reading, yarning, waiting for the pan to boil. After Cluanie in 1954 I was called up for my National Service. I didn't see Elmer again until I met him on the Strathfarrar scheme – I wasn't working there – and by that time he was just about to retire, and I lost touch with him.'

The camp at Cluanie was rough and ready compared with the one at Cannich. 'It was very boggy where the campsite was,' recalled Archie. 'They had to cut a little plantation which was due to be submerged anyway and used the trees as a cushioning in the bog, put tunnel spoil over the top and made the campsite. After about a year or so, the huts started to subside. In Hut 27 when it was raining you had to pull your bed out from the wall or you would get wet. Cannich even had central heating pipes. The heating at Cluanie was by stoves. Jolly cold. I had the flu when I was up there and had to stay in my bed until I recovered enough to go home for a few days. Everything was skimped after Cannich.'

Barry McDermot's first experience of a hydro-electric scheme was the camp near Dalmally that housed the workforce on the Ben Cruachan project: 'There were maybe 600 men in the camp, thirty billets in every hut, and two men in every billet. Wooden huts. They were warm enough, there was central heating and all in them. The sheets were changed once a week. Jimmy Tango, a wee fellow from the Gorbals, was in charge of the hut I was in; he got his name because he was always dancing about. I got a lift up from Glasgow when I started and got booked into the camp. Even if you didn't get booked in, you could hang about and if you knew somebody who was out on a shift you could use his bed. There was a lot of doubling up in the beds, but you still needed a ticket for your meals. I'd never seen anything like this before. Rough and ready, and there were men coming in from the work and going straight to bed, and getting up again and going off in the morning – they wouldn't change their clothes for a week or maybe two weeks. There were no baths, it was all showers that was in it. There was all that muck coming in from the tunnel, and oilskins, and you didn't shave for a week.'

The engineers and higher-ranking staff usually had their own accommodation, a feature some workers recalled as being like the separation of officers and men during the War. At Cannich, some engineers lived in Corriemonie Lodge and had a water turbine for power. The middle-ranking engineers stayed in a wooden hut and had their own club, nicknamed the gin palace. Some rented houses locally. Archie Chisholm's boss lodged with a family in Culligran at Struy. Some of the young engineers were unused to life in the rural Highlands, as Mairi Stewart remembers: 'When the engineers went up first to Cannich, they had to take lodgings in the hotel at Tomich. Two young men, on their first night there, were talking when a third fellow suddenly shouts down. Up the two went, and here was the other fellow – he had got into his pyjamas and a mouse had run up the leg of his trousers, and he had his hands clamped to stop it going higher. The other two, in hoots of laughter, opened the window, put his leg out, and shook until the mouse fell out.'

The camp canteen provided breakfast, dinner and packed meals. As the men moved from camp to camp, they carried with them the memories of the food in any single place. 'They weren't so bad,' recalled Patrick McGinley. 'At that time you were young and, no matter where you might have been, you wouldn't have got enough to eat. There was good food at Whatlings at Invergarry. There was a cook there, Mike O'Toole was his name, and, my God, he was a rare cook – that was great food. Wimpey wasn't too bad.'

'This camp at Cannich was excellent,' said Hugh McCorriston. 'It was the Highland Catering Company. Excellent meals, always plenty to eat. Breakfast, packed lunch, or if you happened to wake up in the middle of the day you could go down and get soup and sandwiches – though it wasn't very often you wakened up – maybe in the hot weather with the heat on the corrugated sheds.'

In November 1948, however, there was a brief strike over food and what was described as 'other camp matters'. The men held a meeting on a Wednesday evening and the night shift decided to down tools; at a second meeting on the Thursday morning, they voted to stay out. One of the strikers told a reporter from the *Inverness Courier* that there had been dissatisfaction for some time: a canteen worker who had given a man a second helping of porridge was to be transferred against her will to another camp, and this petty incident had brought discontent to a head. How long this little dispute lasted is uncertain: it was probably not for more than a day or two but, while it lasted, 1,200 men had shown solidarity.[1]

Mrs Jean Macleod worked in a camp canteen. 'There was just myself and this lassie, Chrissie,' she said. 'And about six or eight cooks who worked on shifts. One was a lassie, a displaced person, Freya, whose brother or cousin, Hans, was with her on site. Myself and Chrissie, we made our quota of sandwiches for the lunch packs. A man got five slices of bread for his piece, with meat or cheese, and one with jam. They called us Thunder and Lightning, because Chrissie put on the margarine like thunder and I would scrape it off like lightning, and we were that quick.'

The lunch packs were issued in paper bags, along with a spoonful of tea and some sugar and, if you were lucky, an apple, an orange or a duck egg. 'I've never seen so many duck eggs in my life. A lot of farmers were on a good number supplying duck eggs. The midday break, knock-off time, was signalled by a hooter,' said Donald MacLeod. 'A teaboy made tea on a brazier-type thing, just a galvanised bucket. It was pretty rough tea, with plenty of milk and sugar in it. You were never quite sure where the water came from or what

Plate 75.
Glascarnoch: lining the main surge shaft with a climbing shutter and a hopper of concrete, March 1957 (*NOSHEB*).

was in the tea, you just took it. Half an hour for midday break. There was a morning and afternoon break as well, ten minutes, for a cup of tea.'

For all his willingness to eat anything, Patrick McGinley's memories of the lunches are not always complimentary: 'Jam sandwiches, Spam, that was about all you got, four or five slices, wrapped up in a bit of paper. There was a "canteen", where a couple of men worked a big boiler to make the tea. You had your tea caddy with you and you'd put it under the tea drum. There was sugar and milk there. You had a break at ten o'clock, another at twelve or half-twelve, and again at three o'clock, you'd get maybe another ten minutes.'

The day-shift workers knocked off in the evening and spilled out from the buses that took them down to the camp, more than ready for their dinner. The queues rapidly formed, each man with his knife, spoon and fork in his pocket, each man thinking that maybe tonight he would be successful in getting a rare second helping. Donald MacLeod remembers the meals as being good, wholesome stuff – soup and mince – but menus had a habit of featuring less appetising dishes with monotonous regularity: 'There was an awful lot of tripe which I didn't like; the sight of it used to put me off. I used to dread looking at all these wash places loaded with all this horrible white tripe. I never did care for tripe and I don't think I've eaten tripe since.

'I never took breakfast there because it wasn't worth going for. What we used to do, this chap from Fort William and myself, after we managed to get a cubicle, was buy a dozen eggs and have two or three eggs in a big jug, topped up with lemonade, in the morning. Raw eggs. It was lovely. Give a good whisk and down that before we went to work in the morning.'

Paddy Paterson helped with the catering in a number of camps – Fannich, Glascarnoch, Glenmoriston – where his parents ran the operation: 'My parents started catering on roadworks on a small scale and built up to the big ones. On the Fannich scheme they took on directly the contract from the Hydro Board. We had the shop and the wet and dry canteens. It was great fun. The raw materials came from William Low and Lipton, and we used the local butcher in Dingwall. Rationing was still in force but it was never a problem in the north of Scotland compared to other places. We provided breakfasts, packed lunches and dinners, and there were cafeterias as well. They weren't open all the time. Breakfast started at six in the morning. The meals depended on when the boys were going out. In Fannich there were three eight-hour shifts – day shift, back shift and night shift – and in other places there were two twelve-hour shifts. Porridge was always a number one in the camps, and the bacon and egg, scrambled egg, whatever. There was no choice, because of rationing, and these fellows were hungry. For the dinners,

28/102 Glascarnoch : Luichart : Torrachilty Project 27/3/57
Glascarnoch Surge Shaft
Lining of main shaft in progress.

Plate 76.
Top. Excavating the foundations for the Glascarnoch dam, March 1954 (*NOSHEB*).

Plate 77.
Bottom. Diverting the main road to make room for the Glascarnoch reservoir, April 1953. In the middle distance stands the Aultguish Inn (*NOSHEB*).

Plate 78.
Opposite. Glascarnoch: building the intake tower (*NOSHEB*).

soup and stews were the main things – good, substantial homemade food. The Poles, Lithuanians, Irish – they all liked that sort of thing.

'The kitchens were all modern, electrified, and some had diesel burners. There was no scarcity of fuel. I reckon we had about forty or fifty in the staff, mostly female, apart from the porter and the chef. Our staff were both local and from further afield, and they lived mainly on the premises, in special quarters. The chips had to be all hand done in those days. We had a thousand mouths to feed. You're talking about forty gallons of soup. A local baker supplied the bread – loaves by the dozen. Milk came in by the churn. A sum was deducted at source from the men's wages for their keep. It ranged from 25 to 30 shillings a week. That got them all their food.'

Life in the camps revolved around sleeping and eating but there were diversions laid on to provide a modicum of entertainment during the off-duty hours. In keeping with the almost-all-male environment the entertainment could be pretty rough. For a start, there were usually wet and dry canteens, the former liberally stocked with bottled beer. The camps on the Logan work sites were run by Willie Logan's wife, Helen, and were generally 'dry', in keeping with the boss's teetotal principles; some drinking still went on but not with the contractor's consent.

'The beer came from various sources,' said Paddy Paterson. 'We had Scottish beers, mainly from Alloa, and Guinness of course. It was all bottled beer in those days. A typical day's beer consumption was colossal. Some of the Irish fellows would go a dozen bottles in a session. There was one guy who normally drank twenty-four in a sitting. And of course there was Carlsberg Special Brew – the one with the kick, and there was the Prestonpans Wee Heavy, and another with a kick was the Double Century. It was cheap – we're talking about pennies.' The dry canteen catered for the more temperance-minded and served only tea, coffee and soft drinks.

The large numbers of men, with pay packets bulking out their pockets, attracted a camp following of less savoury characters. 'We'd come in on the bus from our shift,' said Donald MacLeod. 'As it was Friday, pay night, we finished kind o' early. Out on the square where the buses stopped, where we disembarked, there was a table set out – some Irish ran the Crown and Anchor – with the board. I've known many who would go to the table, put their wage packet down, and on Monday morning they would be at the office getting a sub – they'd lost the lot. I tried betting once or twice but I wasn't winning anything so I decided I wasn't going to lose more.'

Patrick McBride remembers the boys who ran the Crown and Anchor tables as appearing at weekends for a bit of gambling – Toss the Penny and poker as well as the board and dice. 'They would get a kip somewhere for the night', said Patrick, 'and they'd be away on Monday morning. You wouldn't see them again – they'd be at some other camp the next week. That's the kind of characters we had.'

Crown and Anchor was an old game, a kind of poor man's roulette, in which the player betted on the dice turning up symbols to match the six displayed on the board – a heart, a club, a spade, a diamond, a crown and an anchor. 'You've got six squares on the board,' said Barry McDermot. 'You could play it with three dice or six dice, with the symbols of the six squares. You shot the dice and, if four dice came up as some square, that was paid four to one. If only one came up, you lost. You placed your money on a symbol before the dice were thrown. At that time there was so much money floating about the camp, in fivers and tenners. If you put a tenner on the crown, the man who was running the board shook the dice in a box, set them down and lifted the top off to display what they were. You needed two of your symbol to come up to get you two to one, three to get three to one, and so on, and if the whole six came up you got five to one and your money back. Sixty guys could be playing at once, and you'd be standing back. You could put your money on the crown and another guy could do the same. Everybody knew where they had left their money. You didn't touch the board after the dice were rattled. You could have two or three winners in the one game and the man running the game paid out.

'It was the same with the cards. Money up front. No money, no play. It was pontoon mostly; I did see a few brag schools but no poker schools. The dealer had a mate, a teller, who looked after the bets. There would be maybe ten playing, maybe fifteen, in a circle. The dealer watched the players so they wouldn't switch cards or cheat, and the teller watched the money. You weren't allowed to lift your cards away from the table. That was the crack on a Friday, Saturday, Sunday night. That was when you got a break.'

Some of the Displaced Persons, now with large sums of money for the first time, bought new suits and were tempted to gamble: 'They lost everything and they were going around selling suits trying to get the money back,' said Otton Stainke.

Sandy Payne classed his fellow workers into two groups: 'Those who worked, ate, slept, and wouldn't touch gambling because it was a waste of money; and those who made masses of money and just spent it on drink and gambling. My memory of Crown and Anchor is that somebody would come out from Inverness and, with associates he had in the camp, go through to the toilets, take a door off one of the latrines, set it across the wash-hand basins, and lay out the cloth with the "board". Easy to get rid of. Folk just drifted in. It was illegal because the odds were too much in favour of the banker and also because Mitchell [the contractor] didn't want gambling in the camp. To encourage people to play, the banker was willing to give twenty-five shillings in exchange for a £1 note. They hoped to get you hooked.' The resident camp policeman, an ordinary constable, would do a raid on the toilets every so often but could never completely stamp out the games of chance.

'There were fly guys who didn't need to work,' said Donald MacLeod. 'They used to go around all the camps. I was one of the victims on one occasion. I was born and brought up so that when I started earning I would send money home to my parents to help them, which I did. But one day when I came back from work, I found my locker broken in and whatever money I had was gone. These fly guys – that was all they did, they would have a good picking when the place was almost deserted. They would do the Calvine camp today, somewhere else tomorrow, you know. The fly guys – that was all they did – there was no check, you could go into any camp, no security, the huts were wide open.'[2]

Many of the men sent money home every week. In Cannich, Archie Chisholm was entrusted with this task on behalf of some of his workmates: 'I got the job of taking their wee pink forms for their money to the post office for sending away on the Friday night. It was all money orders in those days [the late 1940s]. I can recall now that they'd be sending home £7, that would do their family for the week, that was the average. If they sent £10, that was a really good week. The tunnel was on bonus – I suppose £15-20 was an excellent wage, if you worked what they called "ghosters", double shifts. We used to work ghosters as well, for example if there was a machine broken down and they were desperate to get it fixed. You had to work all night and next day. With a ghoster the wage was good. I was taking home about £6 a week. I started off at £4 10s a week as an apprentice, and rose to about £6. It was good money. The wage on the estate when I'd been working with my father had been £2 10s. You could pick up a pair of boots for 6s. You could go to the town on a Saturday with 5s. Just after the War, spirits were hardly available and a pint of beer would have been 9d or 10d.'

At Dalmally, Barry McDermot knew an Irishman who didn't drink or smoke and bought only a tube of toothpaste every week: 'He was banking all his money, and he was there for about two years, so I reckon he must've saved up four or five thousand pounds, and he was killed out on the job. A lot of men sent money back to their families – my brother-in-law used to send money back to my sister. She sent him a telegram – no need to send any,

I've plenty here, and he showed the letter to a wee Aberdeen man, and the Aberdonian says I wish I had a wife like that, mine canna get enough aff me.'

Many of the workers remember a great camaraderie, especially among the members of the same shift. After he left the Cruachan scheme, a road accident landed Barry McDermot in hospital in Oban with a badly broken leg and, when his former mates in the tunnels heard about his misfortune, they visited him and gave him some cash. 'I was better off than when I was working. There were very decent people up there. You were all mates. If your mate had no money you helped him out, and he returned the compliment. You had to depend on each other. There was no thieving or any of that crack, you could leave your money sitting on the counter [in the bar] and it would be there when you came back.'

'They were great guys,' said Paddy Paterson. 'Salt of the earth, really and truly. Everybody fitted in – they had to, in those circumstances – and they were all mates, no two doubts about it. Sometimes there was a bit of trouble among themselves but there was never anything really violent.' Sir Duncan Michael saw an *esprit de corps* among the members of a team but preferred to describe the spirit among the larger body of workers in terms of adhering to unwritten but totally understood protocols.

'You could be on the job for weeks or months and you wouldn't know [if a man] was a Hugh, a Mick or a John, there was a nickname on them all – like "Mr So-and-so", "May Morning",' said Patrick McBride. 'There was one fellow, Big John, and John was a very early riser in the morning – he would be at the canteen in the morning before it was open, and he had plenty time to walk, but all us young fellows would be running at the last minute and they all knew John so well and would be asking "John, what's the breakfast like today?" Everyone asked him that question. John would look at them and say "I could put the lot in my mouth and say good morning to you".'

Some of the men became hard drinkers. '£48 a week in my hand, after the camp was paid for – the camp was about £4 and 5s or 10s,' said Barry McDermot about the Ben Cruachan scheme in the 1960s. 'If you had a pound at that time, you didn't care who you met. You had enough to buy a round of drinks. It was a lot of money for a young person; I was about twenty. It put people on the drink, they never knew wages like it. £5 would last you all night. A pint of beer was 1s 2d.'

In his memoir of the camps at Dalcroy and Pitlochry, Patrick Campbell writes of the ill-feeling between Scots and Irish and between Protestants and Catholics, but others' memories of the time present the opposite impression. 'There was never trouble between Catholics and Protestants, and I'm talking through experience,' said Paddy Paterson. Barry McDermot said that the Protestants and the Catholics were all the same at Dalmally: 'There was no problem up there. Nothing like that at all. It was only when you came to Glasgow that you met that problem, with the Orangemen and the Celtic supporters. It was all through drink. In the camp, there was no bother. If anybody supported Rangers, they supported them; if anybody supported Celtic, they supported them. I used to come down for some games at Celtic Park. There was no [sectarianism] in the camp.' The curious thing Patrick McBride noticed at Pitlochry was that when the men all got together they got on far better with each other than when there were small groups: 'You could call an Englishman a lazy so-and-so, and it was a laugh and a joke all the time, and he would call you an effing Paddy or something, and it meant nothing.'

At weekends, though, tempers could flare quickly and men would go for each other, usually with fists but occasionally with a weapon. The police might appear with the Black Maria on a Sunday night and by Monday morning half a squad might be missing, dragged away to the cells. In this situation, the camp boss, if someone had this thankless and risky task, might think it best to bend his elbow with the rest of them and say to pot with everything else. It was often beyond the power of a camp boss, on his own, to discipline some 600 men; all he could do was say 'Pack your bags', and the miscreant would just go down the road a few miles to another camp, and he might be back next week.

'Every single weekend there were people thrown out of the camp,' remembered Donald MacLeod. 'A rough crowd. They used to start off in the wet canteen on a Friday night, then on Saturday and Sunday, maybe spend the whole weekend in the canteen. I've seen boys coming in with two or three cases of beer on a Saturday night. Double Diamond, whatever. When the canteen was closed, a party would go on in somebody's hut. It was a tough life but I enjoyed it. You were mixing with all kinds of people and it gave you an experience that otherwise you wouldn't have had. It gave you confidence. It was an entirely different world, with a lot of different creeds thrown together. That's all there was – work all week, and this period at the weekend, and everybody in the same boat. I met people from all over – from the Ukraine, Germany, an awful lot of Poles, some good ones, some bad ones, Irish, an awful lot of Irish, good and bad amongst them, you know. There was a lot of travelling people, except they weren't called travelling people at that time. I knew quite a few of them from my younger days, from travelling around the villages in the islands. And there were quite a few of us from Harris.'

At Tummel Bridge the water distillation apparatus once went missing. It had been borrowed by a guy who made moonshine in the woods, the first stage of which was to steep prunes, potato peelings and other vegetable matter in a big zinc bath.

The more sober workmen kept their heads down. Donald MacLeod and some of his mates from Harris arrived in Tummel Bridge in mid-summer, only a few weeks before the works ran down for the holidays. 'There was three or four of us from the islands and we were moaning the fact we were here just a matter of weeks and we were having to go home. However, we got work, we didn't have to go home after all. There was only three or four of us in this hut. This weekend, this guy, who wasn't a member of our billet, came in drunk. We were in our beds, he was yapping away, somebody shouted to him – the hut was sixty feet long – put the lights out when you go. He had a bottle in his hand. I don't know how many light bulbs there were from end to end of the hut but he went and smashed them all with the bottle, from end to end – that was his "lights out".

'Donald MacLean, from Fort William, and I were in this cubicle one night, both in bed, when who burst in the door but this Sweeney, drunk as a lord, with a knife in his hand. He'd been involved in a fight and he was losing it. In he came. Donald MacLean was a young chap, around about my own age but twice my size, a big, hefty, sturdy chap he was. Sweeney wanted Donald to get up and help him out in this fight. He was fighting more Irish and he'd come into our place for assistance. We refused. The knife was out. "Are ye getting up? Are ye getting up?" The knife was poised. I was never so scared in all my life. We managed to pacify him. It was situations like that, you never knew the minute. Normally that guy was quite placid. It was the drink.'

Quarrels and disagreements on the job, or a grudge nursed since some slight and saved for the release valve of beer at the weekend, lay behind many fights. On the other hand, if a man grew fed up he could just leave quietly and find work somewhere else. There was plenty available and, as long as a man could wield a pick and shovel, he could find a place.

Many of the camps had entertainment of a more edifying sort than a counter serving beer. 'At Cannich, we had the concert hall,' said Hugh McCorriston. 'That's now the village hall I look after to this day. We had entertainment every night. The Archie MacCullough Show used to come up from Glasgow, real professional stuff, and they used to have a big country dance every week, and the cinema three or four nights a week, from the Highlands and Islands Film Guild. There was a concert on either a Friday or a Sunday night. The projection room is still there. Occasionally there would be a really big dance on a Friday night, and of course you had the beer canteen on one side and the dry canteen on the other side, and you never had to go outside. No spirits were allowed to be sold on the scheme – you had to go up to the Glen Affric Hotel if you wanted spirits. You were getting a good wage at that time, and there was nowhere to spend it. You were in the camp and that was it. I didn't drink much and didn't spend too much time in the canteens. The tunnel men – they used to drink a lot, they were heavy drinkers, on bonus all the time.'

Some of the big names in popular entertainment played the camp at Cannich. Sybil Davidson and Mairi Stewart have many happy memories of the dances. 'They had great nights, and the local people took the chance to see stars who would never have otherwise come to their locality,' said Archie Chisholm. 'The first time I saw Robert Wilson was in Cannich hall. You couldn't get in on the night Tessie O'Shea was there. She was a comedienne, a big stout woman, and she sang as well. I think she must have worked with ENSA during the War. Dance bands came too. When there were dances in Struy and Drumnadrochit, there would be buses down from Cannich, and always mayhem, always fights. There were always two policemen on duty and unofficial bouncers, the chuckers-out. The hall at Struy used to be bulging at the seams with all those folk coming down from Cannich. It was a whole new life starting in the glen.'

More unusually, a camp boss at Pitlochry had the bright idea of offering monthly prizes for the best-kept hut surroundings. The number of men who went out in the evening to tidy up and try to beautify the bare, beaten ground with a few ornaments or by planting some flowers came as a surprise.

Some men held on to the spiritual side of their lives. Donald MacLeod remembered men who were more religious than he was: 'You'd see them in the morning, opening the lockers and reading a verse of the Bible before going to work. That was it. They didn't expose themselves as being so Christian or so religious, and they never interfered. You couldn't impose yourself. They were never picked on. Of course they had a job during the drunken spells but again they just kept out of it in their own way without comment. If you did comment, you were in for it.'

The men who came from the Republic of Ireland were by and large Catholic. At Pitlochry some of them built a chapel in their spare time. 'It was done with voluntary labour,' recalled Patrick McBride. 'My uncle more or less started it off. We went to a few other camps – there were a lot of Irish people there – and we could always get a few bob off them on pay day and buy some of the material. The company was very good to us, they gave us a lot of assistance, till we got the wee church built. I forget the day it opened but it was called St Bride's. In fact it burned down years after but it was rebuilt in brick. Originally it had been all timber on a concrete foundation and it took us a couple of years of spare time to get it up.' A priest was appointed to St Bride's on a permanent basis although the workmen had almost completed their labours by that time and had begun to move away. St Bride's church still stands secluded among trees on a braeside near Loch Faskally.

'Ministers would come and they had services in the camps,' remembered Patrick MacGinley. 'I was in Butterbridge, and my brother was there, and we used to go down to Inveraray – he drove a truck, and on a Sunday he would put a canopy on it and we would go down there to the chapel. Sometimes the Church of Scotland would come around but not as much as the Catholics you know.' The Church of Scotland moved a hut it had been using as a school for traveller children from Aldour to Portnacraig camp, Pitlochry, to house a social club for the men as well as provide a place for worship.[3] The Church of Scotland announced late in 1953 that, in the following summer, it intended to give a chaplain a Land Rover and a caravan for two months to serve three work camps in Inverness-shire and that it hoped to raise funds for two more such mobile units.[4]

Barry McDermot remembers very few of his fellow Irish workers going from Dalmally to the chapel in Oban, although the priest would sometimes appear and try to persuade them to attend to their religious duties. On the other hand, Pat Kennedy recalls the Sunday evening Mass conducted in Lochgilphead by Father Sydney McEwan as being well attended.

At weekends, the men often would leave the camps and seek some recreation in any nearby town or village. 'Every weekend we used to go to Pitlochry,' said Donald MacLeod. 'One of the highlights was half a day in Pitlochry on a Saturday. We used to go to this restaurant, up the stair, order a double bacon and egg. You wouldn't call the King your uncle. Some weekends somebody would organise a bus to Perth. Oh aye, they'd go and do a pub crawl there, and then a dance, and, if they were lucky and didn't miss the bus, come back to the camp at midnight. Up north, we used to go into Dingwall or maybe Inverness for a night out, but usually it was Dingwall, as it was a lot nearer.' Sandy Payne's mates nicknamed Inverness Kathmandu.

'The Irish boys worked hard and played hard,' said Paddy Paterson. 'At the weekend they used to go to Dingwall and Inverness and, as they used to say, drink and women were their main concerns. That is natural when you have men working away from home. Many of the Polish lads married local women and settled down. There was one Irishman, always known as the Black Dog – because he wasn't keen on bathing, but he was a gentleman and a great guy, the life and soul of the place wherever he was. I didn't know of any local prostitutes but it was well known there were some in Inverness.'

William Rosie had an old four-cylinder Morris car – 'she never failed but she didn't go very fast' – and made the round trip, five hours each way, between Cannich and John o'Groats at weekends.

In the late 1940s and early 1950s, life was normally quiet in smaller Highland communities. Although some of the places had had military installations or work camps in their vicinity during the War, very few had had experience of catering for large contingents of civilian workmen, many of non-British or certainly of non-Highland origin, with money to spend and eager to forget the stifling fumes of a tunnel or the numbing cold of pouring concrete in a mountain downpour. The camp at Glascarnoch had the Aultguish Inn close by, and the Glascarnoch dam earned the nickname of 'the whisky dam' from the number of broken whisky bottles thrown into the aggregate.

Pitlochry had only two pubs to entertain the weekend influx from the camps at Clunie and Dalcroy. One was in Fishers Hotel and it was here that Danny Boyle was wont to perform his party trick. He liked his booze and was often broke after the weekend but on Saturday night, on good form with a few pints and rums inside him, he would take off his coat and his cap, put his hand down on the ground, palm upwards, and lift into the air any man who offered to stand on it. The generally genial Danny was known as the strongest man on the schemes.

'We took the bus to Inverness at weekends,' recalled Patrick MacGinley. 'I liked Inverness. We had some good times there. They were building a new bridge across the river the time I was there. [The Ness Bridge, another construction by Willie Logan's men, was built between August 1959 and September 1961.] Once you went over that bridge there was a pub on the left hand side – The Gellions. That was where most of us used to drink. The Catholic chapel was across the bridge from there. Inverness was a nice town, they were nice people, very good to you, and they would help you if they could. Of course, there were a lot of Inverness people working on the dams too. We used to go to the cinema – La Scala or the Playhouse.[5] We used to be keen on the girls too. I had a girl but I forget her name now. You didn't stick to the same one too long.'

Iain MacRae used to drive a bus for men going to Inverness at weekends:

'It was a beautiful Daimler touring bus, it even had antimacassars on the back of the seats – after a fortnight you should have seen them. Saturday midday I drove to the town. The bus was free. There used to be a Pakistani who came round every Friday selling clothes. The men bought suits from him, got all spruced up to go to Inverness, and then when I picked them up on Monday morning they never bothered changing, they went straight from the bus into the tunnel, and bought another suit the following weekend.'

'The nearest town to the Butterbridge camp was Arrochar, down over the Rest and Be Thankful,' said Paddy Boyle. 'MacBrayne's bus on Sunday evening at 4.30 up to the camp. We had a long weekend in Glasgow now and then. I stayed with a landlady in digs, most of the lads did that. The family I stayed with was from Donegal – they knew my mother and father at home, and they took me in. I was courting my wife then in Glasgow. I was only about eighteen at the time. I would come up to Glasgow on the Saturday morning and be away back again on the Sunday evening.'

After his first two weeks as a tunneller, Wodek Majewski and some pals went to Inverness on a trip they were to repeat many times. To cover the thirty-odd miles, they took the bus. On the first trip, because they had some money and felt like showing off, they hired a taxi to bring them the last fifteen miles from Beauly to the town. After a drink in the Market Bar or the North Bar, the second stop was the Rendezvous Restaurant on the corner beside the Ness Bridge or the Ness Café. 'At that time everything was rationed but as soon as we got there they said "Oh you're from the tunnel" and gave us anything we wanted – chips, double eggs, bacon. They knew we were the tigers. If there was any money, we went to the pictures, or to the Caley Hotel to the dancing.'

The engineers on the Quoich dam stayed five miles downstream in a camp at Inchlaggan. 'It was very comfortable,' said Laurie Donald. 'We had a cook. There were bungalows for the more senior married engineers on the other side of the road. Our camp was called the Ranch. We did our own messing but the cook was paid for by the contractor. Ten of us lived in the mess, two of them were fairly senior engineers. It was quite wild. We used to have parties. Maybe a couple of guys had cars. No television. Saturday afternoons we used to pile into a works van for a trip to Inverness. Double egg, steak and chips at the Carlton – that was the mecca in those days for a good feed – in Inglis Street.'

'I lived in the camp down in Glenmoriston,' said Roy Macintyre. 'We had a little wing for the engineers. I was beginning to be an old hand then, it was my third year on hydro schemes. This week they decided the Friday night party was going to be in my room, because of the rotation. In the course of somebody attempting to stand on their head on my bed they went backwards through the partition and left a big hole in it. I was feeling worried about having to explain this, but one of the joiners came and put a patch over it in the morning. Generally life in the camp was good and on a Friday night we would go dancing – from Strathfarrar we went far and wide. In Strathfarrar I was with the consulting engineers and we had a nice set up there. Mitchell's camp. We would go to hops in Kiltarlity, Cannich, all of these places; usually the fellows were down one side [of the hall] and the girls down the other and whenever the dances were announced you made a beeline for the one you fancied and usually there were another two guys on a collision course with you. I'm sure the girls loved the dances as well, with the injection of new male talent.'

When a new van was delivered by a big guy called Duncan to the site at Clunie, he and two electricians, new apprentices, drove up to the Tummel Hotel to celebrate. Coming back to the camp, Duncan lost control and the van went off the road. It somersaulted down the brae but when it came to rest one of the apprentices managed to worm his way out to alert the men in the camp. There was no way they could allow the boss to know what had happened – it would have been the sack for Duncan. Half a dozen of the workmen went up to the scene and lifted the van off the big driver. He was unhurt but he was so mad with drink that they had quite a struggle to get him back to the hut and, with him knocking away his would-be helpers, pour black tea into him in an attempt to induce a modicum of sobriety. The demise of the new van was explained away as an accident.

On one occasion the police chased a workman suspected of drunk driving from Drumnadrochit west towards Cannich. The fugitive tore up the road past the camp, bounced down to the dam at Mullardoch, crossed the water in front of the concrete structure, splashed up onto the spoil on the other side and ran his car to the end of the level work area. There he placed a stick to hold the accelerator down, shoved the car into gear and ran it into the loch. His profligacy was explained by the fact that he was earning a small fortune by playing music in pubs and losing the car meant nothing to him. As far as anyone knows, the car lies to this day under spoil in the loch.

The local police forces had foreseen that the influx of workers would mean more for them to do. In October 1946, the Inverness County Police Committee considered how many more constables would be required for Glen Affric. The Hydro Board accepted responsibility for the extra cost.[6] The Chief Constable, William Fraser, in his annual report for 1948, noted how the Cannich population had risen to 2,000 and was expected to reach 2,500; he added that people 'can truly be said to have gathered from all corners of the earth'. Cannich's crime rate had run at about twelve cases a year in the old pre-hydro days, and there had been only four cases in 1946 (one of theft, two offences against the Dogs Act, and one involving a motor vehicle). In 1947 the number of cases rose to sixty-six, and rocketed to 198 in the following year, comprising seventy-two crimes against property, forty-five breaches of the peace and six crimes against the person.[7] In the discussion following Fraser's report, Lord Lovat paid tribute to the workers who, he said, 'did not indulge in poaching'. The number of petty assaults and breaches of the peace in Inverness-shire rose from 153 to 211 in 1952, chiefly because of the workmen at Loch Quoich and Invergarry, reported the Chief Constable.[8] However, his northern colleague, in Ross and Cromarty, where reported offences rose by only 5 per cent in the same period, ascribed no blame to the work camps.[9]

A trickle of offenders from the camps added some colour to the roster in Inverness Sheriff Court. A Glasgow labourer was sentenced to twenty-one days for stealing an electric guitar from the Cannich camp cinema and sending it, wrapped in two British Rail carriage blinds, also stolen, to his girlfriend in the south.[10] Three cases were dealt with early in December 1949. When Thomas Jamieson, a ganger from Auchtermuchty, pled guilty to fighting with a foreman in the Affric tunnel and attacking a locomotive driver, a breach of the peace for which he was fined £10, his solicitor stated that this was a common case for Cannich. John Watt Bruce of Gardenstown got into a fight in the dry canteen in defence of the reputation of two women and, as no drink was involved, was fined only £1. Ernest Stewart, from Kincardine O'Neil, was fined £5 for borrowing a jeep without permission to drive from Cozac to Cannich, and the Sheriff added another £1 fine as Stewart did not have third-party insurance.[11] Two months later, however, fines were considered not sufficient punishment for the Lerwick man, James Jameson, and his Dundee pal, James Macdonald, who broke windows and caused trouble at a dance in Beauly, and both were sent to reflect on their behaviour for ten days in jail.[12]

During the construction of the schemes at Loch Awe in the 1960s, Oban became the main weekend destination, where one of the attractions was the dancing in the Corran Hall. It was, however, twenty-two miles to the town, a long way to hitchhike back in the winter if you missed the last bus, and many often preferred to seek entertainment in the villages closer to the site. Five workers – four of them Irish and the fifth an Argyllshire man – were charged with breach of the peace at Oban Sheriff Court in January 1962 after fighting at a dance at Taynuilt: 'Undaunted by the odds, burly constable John Munro drew his baton and with the aid of an off-duty constable and two civilians quelled the disturbance,' commented the *Oban Times*, adding that the procurator fiscal considered the brawl 'quite the worst' since the schemes had started. The five miscreants were fined a total of £130.[13]

Longer holidays were scarce for most of the workmen. 'When the Fair holidays came up, you'd get two weeks,' said Patrick McBride. 'You wouldn't get any other holidays, except maybe a long weekend. But you didn't get as many long weekends at that time. You'd be off a couple of days at Christmas, maybe a week. You could come home [to Ireland] but be back again before the New Year. At that time you got the owld boats from the Broomielaw to Derry. A very slow system it was. Just all night. It would be six o'clock in the morning when you got to Derry, and everybody sick – it was a terrible journey. And the same going back. They carried a lot of cattle on the boat at that time, going back, and they were all down below, and their breath coming up. We used to catch a bus from Derry right to Creeslough, about forty-odd mile. The bus stopped everywhere. You left Derry at five o'clock in the evening but you might not get here until seven or half past seven. Then we had to walk a mile to the house. I remember going away sometimes when you had no suitcase but [carrying your belongings] in brown paper and string – and by the time you would get to the bus it was all wet. In Derry, there, we used to go up to Woolworths, see the lovely cases and try to save half a crown for one.'

Patrick MacGinley came down to Glasgow infrequently for a weekend, or spent a couple of nights in Inverness. A death in the family might bring a man home to Ireland for a longer period of time but generally they preferred to work, transmitting money back to their families.

On Boxing Day 1947, Patrick Gilmore, a native of Galway but then resident in Inverness, was setting sticks of gelignite in rocks at Mullardoch when one charge exploded prematurely. Gilmore was killed outright by the blast. The only eyewitness explained that Gilmore had been wiring up the charge but a later examination of the cables showed that the detonator had not been connected at the time. The jury at the public inquiry into the accident agreed with the view that no evidence remained to explain the cause of the explosion.[14] Gilmore's death was the first of several on the Glen Affric project.

Early on the morning of 2 November 1951, a twenty-six-year-old tunneller from Strabane, John Haughney, was fatally injured one mile inside

Plate 79.
Excavating a cut-off across the Glascarnoch road diversion, August 1953 (*NOSHEB*).

Plate 80.
Filling the temporary openings at the base of the Glascarnoch dam, April 1957 (*NOSHEB*).

the Errochty tunnel in a sudden explosion at the drill face. One of his companions, Hugh Friel from Kindrum, Donegal, was less seriously hurt and taken to Bridge of Earn hospital. The other five drillers on the shift suffered cuts and bruises. The cause of the explosion was suspected to be accidental contact between the drill bit and a misfired detonator left in the rock face; and this suspicion was upheld at the inquiry held a week later. A witness, John Broderick, testified that a red ring was usually painted around a hole where a misfire had happened to warn drillers to stay clear of it. The shift boss, Walter McLean, speculated that Haughney's drill tip may have slipped across the face and lodged in the old hole.[15]

Injuries and deaths happened on almost all the hydro-electric schemes; the building of Glascarnoch dam was among the few exceptions where there were no fatalities.[16] Six drillers were injured in a blast in the penstock tunnel between Comrie and St Fillans in December 1955.[17] The names of the five men killed in the construction of the Clunie tunnel are remembered on a plaque on the Clunie Arch spanning the road down to the Clunie power

Plate 81.
The Glascarnoch dam nearing completion, June 1957 (*NOSHEB*).

station; many power stations have memorials of this sort. Usually the mates of a casualty mourned in their private ways and then the work went on.

'There was no such thing as counselling,' said Archie Chisholm. 'There was a feeling that if you had a narrow escape you acknowledged it and got on with the job. If someone was killed, you had to put it at the back of your mind.'

'The tunnel at Glen Errochty collapsed, the whole thing collapsed, one or two boys were killed,' recalled Donald MacLeod. 'Mind you, there wasn't all that many fatalities, considering the sort of equipment and the amount of men. When you think back, there wasn't even a medical attendant on site. I had to go to Pitlochry when a flake from new corrugated iron went in my eye. I went in to the timekeepers' office and they had a look, but they were only timekeepers. They laid on a car to take me to the doctor in Pitlochry, and he took off the flake with a magnet. Then it was straight back to work again.'

In the early days, health and safety standards were lower than they became, and were often largely ignored. Every man had to look out for

Plate 82.
The buttress wall at the Monar dam when freezing air stopped concreting, 5 December 1961 (*Press and Journal*).

himself and use his common sense. Even when hard hats became available, many chose not to wear them – one man told me they were like paper anyway. Two men at Trinafour were drowned in the fast-flowing water below the dam, when the first man fell in and his mate tried in vain to get him out. 'The ambulance came up but they wouldn't let the bodies be put in it,' recalled Don Smith. 'They were dead when they fished them out, and they said the ambulance was for the living. The bodies were put on a Chevvy and driven down to the shed that served as a mortuary. I was disgusted, black affronted. There was an inquiry nine months or a year later but nothing was ever heard.' An Irish worker, Charles Donachie, was drowned in a spate in the Affric river in January 1949 when he was trying to free flotsam jammed against the uprights of a footbridge; the foreman saw him for the last time when he hit a rock 150 yards downstream and, although his jacket was later found, his body hadn't been recovered by the time of the inquiry.[18]

Experience probably kept many a man alive and whole but even the expert could fall victim to the thoughtless moment or bad luck. A rockfall during the excavation of the surge shaft in the tunnel between Loch Mullardoch and Loch Benevean fractured the skull of Edward MacLean from Lybster and broke James Douglas's jaw. MacLean compounded his injuries by falling one hundred feet down the shaft and later died in hospital.[19] This accident led to criticism of the safety precautions on the scheme but a spokesman for the contractor, John Cochrane and Sons Ltd, said the measures were adequate.[20] In another accident, Balfour Beatty and Co. Ltd and John Bisset and Sons Ltd admitted liability when George Jack from Cullen sued for damages after losing his hearing in one ear and suffering other injuries from an explosion at Fannich; Jack asked for £3,000 but the Dingwall sheriff thought £954 17s 2d a more fitting sum.[21]

A fire broke out in an office in a hut in the Breachclaich camp in the small hours one morning in March 1960. The thick acrid smoke aroused two cooks who awoke the sleeping workmen. Until the fire brigade could arrive from Killin and Comrie, the men saved most of their personal belongings and fought the blaze with buckets and extinguishers but, after three quarters of an hour, there was nothing left of the hut but ashes and charred timber. The contractor, R. J. McLeod, brought in caravans for the men until the job was finished.[22]

Three men lost their lives when two huts in the Glascarnoch work camp caught fire at four in the morning on 27 April 1953. The others in the hut tried to fight the blaze but failed to stop it leaping to other buildings. The Dingwall fire brigade took forty minutes to negotiate the thirty miles of single-track road to the scene. Some workers lost all their possessions but

Plate 83.
Winter – erecting the Monar dam, 5 December 1961 (*Press and Journal*).

others managed to stagger out with their belongings to the safety of the braes.[23] 'That was a tragedy,' said Paddy Paterson. 'My brother had a place in that accommodation but he managed to get out of the window. It was caused by an electrical fault. It would have been worse but for a fellow who got a digger and took the end off the hut to let the men out.'

The notoriously unpredictable Highland weather caused some accidents and a few deaths. In June 1953, on Coronation Day, it snowed up at the Cluanie dam site and an elderly worker succumbed to hypothermia. More often the weather simply added to the list of discomforts the men had to endure.

Severe frost in the snow-laden winter of 1955 made concreting impossible and brought work to a halt across the Highlands: the Mitchell company in Glen Moriston paid off 600 Irishmen who went home until conditions improved; at Cluanie, where no outside work could be done, one man reported that it had been too cold to sleep in the hut, despite six blankets; Logan paid off 300 men but the tunnellers at Invermoriston kept going, warmed in the depths of the mountain. Some men working on a head pond were cut off for a week and are said to have augmented their supplies with fresh venison from the local herd. The frost lasted for around two weeks.[24]

And of course there was the rain – at times it must have seemed incessant. In December 1954, ten inches fell in twenty-two hours at Quoich. 'One month we had more rain than London gets in a year,' said Laurie Donald. But there were also days of brilliant sunshine. On one such day, Archie Chisholm and George Bain were sent to fit a brass flange to a valve on the surge shaft between Fasnakyle and Benevean. It took some time to bore twelve holes in the metal with a hand drill but 'we weren't worried because the weather was so good. We were nearly finished [when] there was this gurgling down below and a fountain of water came up and completely submerged us. The shaft was designed for that, of course, but we hadn't banked on it happening while we were up there. We dried ourselves out in the sun.'

[5] '... power and light have come to us ...'

The official opening ceremony of the Loch Sloy scheme took place on 18 October 1950, a damp, cold Wednesday. The Board published a handsome, well-illustrated, hard-covered brochure to mark the occasion, and took the opportunity to blow its own trumpet a little. 'Today', wrote Tom Johnston in the Foreword, 'the spreading distribution in the remote parts of the North, the eagerness of the Highland people to make use of electricity, the new spirit that exists where before there was only depopulation and despair, the establishment of a new industry in Scotland to manufacture machinery for the new power stations, the new employments that have been provided, already supply an effective answer to any who may lack faith in the future of our country.'[1]

The dam on Loch Sloy was a symbol as well as a literally concrete reality. It was touted as a national enterprise involving workers from every region, and equipment and material from every corner of the land. To grab a little bit of historical cachet and tap into Scotland's sense of itself as an ancient, proud nation with a somewhat bloody past, the brochure included some paragraphs about the Clan MacFarlane, on whose ancestral land the scheme stood. 'Loch Sloy' had been the war-cry of the MacFarlanes, it reminded its readers, and the clansmen had often fought in defence of the nation, at Largs, Bannockburn and many another conflict. The moon had been 'picturesquely' nicknamed 'MacFarlane's lantern' in some places, a reminder of the clan's propensity for night-time cattle raids on its neighbours, but, said the brochure, the lantern of the twentieth century was the electric lamp in cottage, hotel and street. In setting the tone for the opening, Sir Walter Scott could not have done better.

Queen Elizabeth (the late Queen Mother) performed the ceremony.[2] A huge crowd had gathered to see Her Majesty and witness the occasion. Tom Johnston had cleverly arranged the installation of an electric bulb under every seat in the stand to take the edge off the chill for the waiting throng, and the band of the Royal Scots was on hand to provide some stirring music.

Speaking from the open-air stand looking out across Inveruglas Bay towards the distant mass of Ben Lomond, Queen Elizabeth paid tribute to the 'considerable hardship' endured by the builders, the 'vision, tenacity and technical skill' of all who had been involved, and said their reward would be the 'new strength surging into the very arteries of Scotland's being'. She earned praise from the Highland newspapers for talking of the 'compelling' need to bring electricity to the villages and glens and stating that 'in Scotland we want our rural areas to be equally flourishing and equally favoured'. When she finished, she pulled a lever to set the generators in motion.

A more humble but perhaps more significant ceremony had already occurred. In April 1948, the village of Arrochar at the head of Loch Long had been electrified by a line from the diesel generator at Loch Sloy. In a gesture that was to be repeated in many another local celebration up and down the Highlands to mark the coming of the 'electric', the switching-on had been done by the oldest inhabitant, 96-year-old Miss Mary MacFarlane.

Less than a year later, the official opening of the Pitlochry power station and the Tummel-Garry scheme was in the final stages of preparation when Sir Edward MacColl, whose wife was to be the principal guest at the ceremony, died suddenly on 15 June 1951. As chief engineer and deputy chairman, Sir Edward (he was knighted in 1949) had been half of the vital leadership – he as the engineer complimenting Tom Johnston's political skills – that gave the Board real drive. He was one of the pioneering geniuses of hydro-electricity in this country.

MacColl was born in Dumbarton in 1882. After an apprenticeship in John Brown's Clydebank shipyard, he joined the Glasgow Corporation Tramways Department as an electrical engineer. In 1919 he became chief engineer with the Clyde Valley Electric Power Company and helped bring into being the hydro-electric scheme at the Falls of Clyde. When this project was complete in 1927 he moved to the Central Electricity Board and created in central Scotland the first regional electricity supply grid in Britain. He introduced to Scotland the concept of pumped storage hydro-electric schemes, based on the principle of reversing the flow of water through a turbine so that, at times of low demand for electricity, water can be pumped from a lower reservoir back up to a higher one, where it can be stored for later release when more electricity is needed. In 1936 he put forward this concept in relation to Loch Sloy and Loch Lomond but it was too far ahead of its time; Loch Sloy was realised, as we have seen, as a conventional scheme, and pumped storage remained on the drawing board for another thirty years.

Some of his colleagues felt that overwork contributed to Sir Edward's

Plate 84.
The Storr Lochs project on Skye. The foundations for the power station and one of the anchor blocks for the steep cable railway being laid (*NOSHEB*).

Plate 85.
The Storr Lochs project on Skye. The cable railway leading down to the power station beside the sea (*NOSHEB*).

death at the age of sixty-nine. J. Guthrie Brown called him 'a bonny fechter' who did not delegate easily[3] and the struggles in the early years of the Board to overcome political opposition, shortages of labour and materials, and the myriad of other problems arising from the Board's development programme probably did take their toll. 'He literally wore himself out ... in the service of his country,' said Tom Johnston.[4]

Instead of a grand opening in the manner of Loch Sloy, Pitlochry was the scene of the unveiling of a bronze plaque to Sir Edward's memory in a subdued ceremony on 14 April 1952. Lord Cooper, who had headed the Committee that had brought the Board into being in 1943, made a speech in which he said:

> Scotland has always been renowned for her engineers but even in that distinguished company the name of Edward MacColl will long be remembered ... We may truly say as we stand here in Pitlochry, within sight of this great engineering enterprise, and with the sound of the harnessed water in our ears, if you would see a monument to Edward MacColl, look around you.[5]

The Board commissioned the building of a number of small hydro-electric schemes to serve some of the more remote areas of the country and the islands. The Morar River and the Allt Gleann Udalain were both harnessed in the first constructional scheme in 1945; and others followed over the next decade. Each produced just a few million units per year but they made a vital difference to the localities they served. One of these, at the Storr Lochs on Skye, was begun in 1949 to serve the whole island. Work was already underway when Alastair Kirk, who, after his two years national service with the Royal Engineers, had newly joined Blyth and Blyth, the long-established Edinburgh firm who were consultants for the scheme, came off the train at Mallaig to board the steamer *Loch Nevis* for the voyage to Portree. It was a long journey – the train had left Waverley at half past five that morning and it would be five o'clock in the afternoon before he would reach the work site seven miles north of Portree. He was due to relieve other engineers, about to go on holiday, but what should have been a four-week stint grew to eighteen months.

The Storr Lochs scheme used two existing lochs – Fada and Leathan – beside the road from Portree to Staffin, below the famous landmark of the Old Man of Storr. A camp was established beside the road, and an access road was built into the dam site, at the north end of Loch Leathan. From the

dam a pipeline plunged four hundred feet down the steep coastal cliff to the power station located virtually on the beach. A cable railway, with a maximum gradient of 1 in 1.8, was put in with a capacity of about ten tons and everything for the power station had to go down it. The Storr Lochs power station is still without road access. Early in the construction, the contractors, James Miller and Partners Ltd, thought they could bring materials in by landing craft but the vessels broached in the strong tides and one was lost in Bearraraig Bay laden with equipment.

'The construction took a long time because of the remoteness,' said Alastair Kirk. 'Everything from the mainland had to come by road via Kyle of Lochalsh. The contractor could get a phone call saying such-and-such had arrived at Kyle, and they had to send a lorry down thirty or forty miles of very narrow road to get it. The MacBrayne ferry at Kyle in those days had a rotating platform on deck. During spring tides it stopped for about two hours in the middle of the day; the Kyle slip was so short then that if the tide fell below the end the vehicle platform on the ferry was not in line with the quay and the ferry couldn't dock. At the start it could take only two cars or one lorry but by the time we finished it could take more. The old road from Kyleakin to Portree wound along the coast. All the materials had to be brought up it, including loads of aggregate from Sconser. Initially the road from Portree to the scheme was not much more than a seven-mile-long track. Towards the end of the construction, when they needed to bring a forty-ton generator to the power station, the driver was taken from Portree to see the road he would have to negotiate: "I'll go but I won't stop at all," he said. "If I go off the road it's too bad, but if I stop I'll just sink" – and that's what he did, he drove it well.'

The sixty-strong labour force were mostly Irish and Lowland Scots – many of the joiners were ex-shipyard workers from the Clyde – and lived in a hutted camp beside the Staffin road on a moor open to howling gales. From the camp they used to walk across the bogs to the site. At weekends they visited the pubs in Portree where the licensing laws put a stop to service at only 9 p.m. The engineers shared a wooden bungalow and had the use of a small BSA Bantam motorbike for transport to the island capital.

The concrete in the fifty-foot-high mass gravity dam had to be tamped in place with the workers' boots. At first a single 42-inch steel pipeline was welded and bedded on concrete stools to feed the power station but a few years later a second pipeline was installed to fill the requirements of the original design. A complicated trifurcation at the foot of the cliff split the water flow to feed three turbines in the power house. The plan to generate with two sets and keep one in reserve had to be revised when demand for electricity on Skye made it necessary to operate all three.

'It was a good job', said Alastair Kirk, 'and gave me a lot of experience. There were no particular problems in building the dam, as the rock was quite good. I remember some of the rock being full of fossils.'

The Storr Lochs scheme was officially opened on 31 May 1952, on a beautifully hot day. After a few months working on other engineering projects in the Kyle area, Alastair Kirk was appointed as the resident engineer on another small scheme – at Loch Dubh, high in the hills on the eastern flank of Strath Kanaird, seven miles north of Ullapool. On this occasion the main contractor was R. J. MacLeod. A novel feature of the scheme was that the main sixty-foot dam on Loch Dubh was built with colcrete rather than concrete. Alastair explained: 'With colcrete you build up a three- to four-foot layer of loose stones, all roughly two to three inches in diameter, and put in grouting pipes. The shuttering has to be very tight to withstand the injection of grout, a colloidal solution made in a special mixer. The grout does not mix with water and pushes any water in the stone out as it is pumped in under pressure and penetrates the interstices. This method was an experiment to a certain extent and it worked out all right.'

The main dam at Loch Dubh fed water to a smaller reservoir behind a twenty-foot intake dam, whence the water flowed through a single 27-inch steel pipeline, two-thirds of a mile long, to the power station on the floor of the strath. The head was high – 540 feet – and fed two 450kW turbines. The electricity served some 600 consumers between Braemore and Stoer, and lines were planned to extend the distribution as far up the west coast as Durness. Strath Kanaird did not have to cope with the supply problems associated with the Storr Lochs and the scheme was completed in record time – at least officially. At the opening ceremony on 4 September 1954, Tom Johnston drew attention to this fact and made a gentle joke at R. J. MacLeod's expense: 'In just under two years from the placing of the contract, this ceremony of completion is being held, possibly because of the personal interest of the contractor, an Ullapool man.'[6] The fact was that the scheme was not completely finished but was able to generate power.

On this scheme the key workers had been housed in caravans, accommodation the men reported as being superior to huts, and it was felt that this had contributed to a drop in turnover in the labour force.

The principal guest at the ceremony was Sir John Cameron, Dean of the Faculty of Advocates and Deputy Chairman of the Highlands and Islands Advisory Panel, who had all those years before chaired the Loch Sloy and Tummel-Garry inquiries. 'Although we are accustomed to think that the age of miracles is past, we are seeing in effect a miracle of faith and hope being accomplished in the harnessing of the waters,' he said. 'It is all very well', he went on, 'for these lovers of the country in the Athenæum and in Pall Mall to talk about desecration of the Highlands by electrical power stations, but how many of these south-country Highlanders would like to get up at six o'clock in the morning and go 200 yards to carry water from a well? How many would like their wives to do that, or trim stinking oil lamps or carry creels of peat on their backs ...?'

Shortly after Sir Edward MacColl's death, Hamish Mackinven joined the Board as press officer. A native of Campbeltown, Mackinven had cut his teeth in journalism on the local paper before serving with the RAF in Burma, and resumed writing on a freelance basis while working as a labourer for the Forestry Commission after the War. His contributions to the left-wing magazine *Forward*, which had been founded by Tom Johnston, brought him to the notice of George Thomson the editor and after a short time he became Thomson's assistant. In the late 1940s, Mackinven worked as press officer at Transport House, the London headquarters of the Labour Party, for some of the leading figures in the Labour administration, including Clement Attlee and Nye Bevan, but he always nursed a wish to return to Scotland and in 1952 he responded successfully to an advertisement from the Hydro-Electric Board.

'I was interviewed by Tom Johnston,' he said. 'I'm the only person alive who actually worked with him on a day-to-day basis. He is maybe the greatest man I've ever met. Handsome, always dressed in the sober, old-fashioned, double-breasted dark blue worsted suit, the virgin white shirt, the handcrafted bow tie, the black nap overcoat with the velour collar and the homburg. He was chairman of the Board but he had no office. He would come in by car from Fintry where he lived to 16 Rothesay Terrace [the Board headquarters in Edinburgh], come up the steps into the hall, open the door of the board room with the twelve-place table, hang up his coat and his long silk scarf in the hall, go to the table and pull out the chairman's chair, lift the old attaché case on to the table, open it and place the travelling clock on one side – and that was the shop open. He would pull out the chair on his right for his secretary who was also head of the typing pool, and another chair for the people who were coming to see him during the day. He would always say to me "Part-time chairmen don't need a room".'

Johnston was a bibliophile and in the Fintry home there was said to be the best privately-owned library on Scottish history ever assembled. This passion for learning and for his country's heritage provided him with a

Plate 86.
Tom Johnston at the opening ceremony at the Lawers power station (*Perth Museum. Copyright: Louis Flood*).

profound knowledge that served him well at the opening ceremonies. Although a modest man and not an accomplished public speaker, he always seemed to come up with the right thing to give each ceremony its distinguishing mark. When the Gaur dam on the edge of Rannoch Moor was opened in June 1953, Johnston invited Robert Menzies, the premier of Australia, in Britain for the Commonwealth Prime Ministers' conference, to perform the ceremony. He knew that the dam stood in the territory of Clan Menzies. As well as this historical link, near the scheme grew one of the last surviving clumps of the Old Caledonian pine forest that had once blanketed much of the Highlands. Hamish Mackinven describes what happened:

> When the ceremony was finished, TJ leaned under the table and took out a brown paper bag and gave it to Menzies. It contained seed from the reseeding of the forest. Menzies was over the moon, took the seed back to Australia and had them planted on the way to Parliament House in Canberra.[7]

Throughout the late 1940s and the 1950s, electricity came to community after community. 'In the village of Tomich in Glen Affric every house is either supplied or ready to be connected,' reported the *Inverness Courier* on 12 October 1948. 'Housewives ... have not been slow to take advantage of the benefits of electricity. Already the latest domestic electric cookers are being used in 20 per cent of the houses ... Other electrical appliances ordered or in use include irons, kettles, immersion heaters, washing machines and refrigerators ... In Cannich the first all-electric hotel in the area is now operating ...' The hydro power station was not yet in operation and the power came from a temporary 3,800 kW diesel generator operated by the Board primarily to supply the work camps and machinery.

In July 1949, Tomintoul, the highest village in Britain, had its switch-on, when the spur line from the main distribution line in Glenlivet was completed. The pylons, strengthened to withstand the winter storms, crossed nine miles of moor at 1,250 feet above sea level.[8] Electricity reached Drumnadrochit in May 1950, and Fort Augustus, Kiltarlity and North Kessock early in 1951. Revd Hugh M. Gillies, the parish minister in Fort Augustus as well as a county councillor, gave an appropriately Biblical flavour to his speech at the opening ceremony: 'We are indeed pleased that after many years of hope deferred, our faith has been rewarded, and power and light have come to us ... We have entered into our true heritage.'[9] The Bunloit switch-on ceremony took place in November 1954, and those at Applecross and Dallas, the former a famously remote district on the west coast of Ross-shire and the latter a village in the more accessible Morayshire hills, in July 1955. In March 1956, a ceremony at Braes on Skye marked the supply of power by undersea cable to the island of Raasay; when the switch was thrown on the mainland, three rockets were fired from the island to mark the arrival of electricity.

In a few communities there were marked signs of impatience for the new energy. The residents of Kilmorack, within sight of the Beauly switching station but still unconnected, protested about this 'flagrant wrong' in May 1951.[10] Kilmorack had to wait a little longer but, at last, on the evening of 4 March 1953, Miss Mary Macrae, aged eighty-four, of Old Post Office House, pressed the button to bathe the village hall dramatically in light. Her words were a gift to the Hydro Board's press officers: she had decided to have her house wired, she said, so that she could enjoy electricity 'during the remainder of her pilgrimage here'.[11] On the following day an exhibition of appliances was mounted in the hall. These exhibitions, along with demonstrations by cooks and sales staff, were a feature of opening ceremonies and were understandably very popular with the public. In June 1947, the Board mounted an exhibition in Fort William that included sheep shears, welding equipment and a grain-crushing plant as well as the more humble kettles and heaters; and the public placed orders to the value of £2,500 in the first two days.[12] The Hydro Board shop, selling appliances and collecting consumers'

Plate 87.

A classic shot of a cooker being delivered to a crofthouse somewhere in the Highlands (*NOSHEB*).

bills, was to become a standard fixture in every high street north of the Highland Line; they sold over £1 million's worth of appliances in 1960.[13] The Board staged an Electrical Economy Exhibition and a display on agriculture in Inverness in April 1952, in collaboration with the quaintly named Electrical Association for Women who hosted a lunch in the Caledonian Hotel on the first day, and welcomed 10,000 visitors in the first week.[14] To promote new equipment for hotels and guest houses in ports between Oban and Ullapool, the Board launched a floating exhibition on a chartered motor yacht, the *Western Isles*, in the autumn of 1964.[15] An exhibition of appliances ran in the Corran Halls, Oban, for three days in May 1965.[16]

Highlanders also had to learn to cope with another new phenomenon, one that was to become a standard occurrence in the winter months – the power cut. From the start, however, the Board's linemen made great efforts to keep the electricity flowing. In January 1952, when a ship at Kyle of Lochalsh dragged her anchor in a storm and snagged the undersea cable to Kyleakin, they seemed determined to be stopped by nothing to fix the fault.

When two men driving with repair equipment became stuck in a snowdrift in Glen Garry they abandoned the vehicle and trudged twenty-three miles to reach Kintail. The engineer on Skye carried an eighteen-foot ladder for half a mile over rough ground to get to the pole at Kyleakin where the cable came ashore. Despite these heroic exploits, the Skye folk remained for fourteen days without power.[17]

In September 1957, Iona was linked to the grid. This was not just another island. After all, as every Scot knew, Iona was in many ways the spiritual heart of the nation – Saint Columba had chosen it as his base for bringing Christianity to the Picts, the Book of Kells had been created here, it had been a beacon of learning during the Dark Ages, and many of the ancient kings of Scotland were buried in the island's sacred soil. With his genius for public relations, Tom Johnston recognised at once that the switch-on ceremony on Iona had to be special. Hamish Mackinven, the Board's press officer, recalls the event:

> Johnston sent out so many invitations that it required special first-class trains from Glasgow and Edinburgh to Oban to carry the guests. I didn't know how I would cope with all these hundreds of people. I said to him I was getting a bit worried. He leant back in his chair for a minute and, I'll always remember this, said 'Hire the *King George V*.' That was the biggest boat MacBraynes had. I rang the chairman of MacBraynes. Of course, the man was bemused but the moment I said Mr Johnston would like it, there was no problem. That's how it was done. Only a very small party went ashore for the actual switch-on. With us were people from all the Scottish papers and from *The Times*. In those days *The Times* was still 'the thunderer', read where it mattered all over the world. On the way back to Oban, Tom Johnston asked me if Mr Brown (*The Times* man in Scotland) was on board. He was, and Johnston spoke to him. On the following day *The Times* published an exclusive story.

The author of *The County of Argyll* noted 'a silent revolution' proceeding with the coming of the power supply.[18] By that time, only a few isolated corners of Argyllshire remained without electricity. As the electric bulbs went on, literally all over the countryside, there were a few moments of reflection. The Revd W. J. Macintyre wrote of the switch-on on Seil in the parish of Kilbrandon: 'On the same day, and perhaps not without a feeling of sadness, the lights of the paraffin lamps on the island went out for ever.' Generally, however, there was rejoicing; life in the countryside, especially in winter, was hard enough. 'The three-minute tea kettle is an especial boon, and the electric radio set an added luxury,' commented the Revd Duncan MacAulay in Inverinate in 1953, before adding 'Many find the installation of electricity within the home a financial burden.' [19]

No doubt many crofters found the wiring up of the house an unplanned-for expense, and this became a cause for complaint. At first any household or premises within half a mile of the Board's distribution lines was connected without charge. Within a few years, however, rising costs forced the Board to adopt a policy whereby new consumers agreed to use or pay for an amount of electricity equivalent to 10 per cent (raised to 12.5 per cent in 1950) of the connection cost. The Local Authorities Committee noted in 1951 that some remote crofters were being asked for as much as £140 to be connected,[20] as the Board sought to recover some of its expenditure by charging line rentals. Sixty people gathered in Balnain school in Glen Urquhart in October 1952 to hear the chairman of the Board's Consultative Council, none other than G. T. McGlashan, now firmly on side with the Board after the dispute over the Tummel-Garry scheme, explain the plan for the electrification of the glen. The Board was asking for £3 per room per year guarantee from potential consumers. Many thought this unreasonable and unfair, as their neighbours down in Drumnadrochit and Milton were being connected free of charge. A crofter with a six-room house said he couldn't afford to buy enough appliances to consume £24 worth of electricity in a year. McGlashan explained that it would cost the Board £24,000 to supply the 139 households in the glen; the Board was seeking a guarantee of only £850, he argued, for what would cost the Board around £4,000 per year to maintain – this represented in real terms a subsidy by the Board.[21]

Sir Hugh Mackenzie, the deputy chairman of the Board, held a series of meetings late in 1953 to explain to the communities scattered along the rugged north coast of Sutherland measures for connection to an interim power supply from a diesel plant in Wick. The people in Melvich, Bettyhill, Melness and Tongue were asked to pay a levy of £1 per room per year on top of the normal tariff to defray the high cost of the installation, reckoned to total £250,000. The maximum any house would have to pay was £10 and this levy would cease as soon as hydro-generated power came on stream to serve the area. The revenue the Board expected to earn from the coastal districts was only £3,300 per year and, explained Sir Hugh, the levy was necessary as the Board received no public subsidy. Most of the potential consumers balked at this outlay and Sir Hugh expressed disappointment.[22]

Shortly afterwards, more accepted the levy, however, and just two months later, in January 1954, Miss Mary MacIntosh the Melvich postmistress did the switch-on in the crowded, decorated village hall. Of the 312 potential consumers, 171 had taken a supply. Sir Hugh Mackenzie reminded his listeners of the cost of bringing electricity to remote places. 'The 'provision to some rural lines', he said, was costing the Board £1,200 per week and was made possible only by the sale of surplus power outwith the Highlands and the re-investment of the profit to meet Highland requirements.[23] As it was, 78 per cent of the Board's output was for consumption in the Highlands and Islands.' Applause greeted his remark that the Board had done as much in six years as others had done elsewhere in a lifetime.[24] By this time consumers between Ullapool and Lochinver had been connected and the line around the coastal districts of Drumbeg and Nedd would be ready by summer. Another line was being extended west from Thurso and the switch-on in Strath Halladale was expected soon. It was to be another six months before the hydro scheme at Loch Shin was to start and another seven years before it began to generate.

Electrification represented a considerable investment by the Board. The bringing of power to the village of Dores on the northern shore of Loch Ness in December 1952 was estimated to have represented a Board subsidy in the area of £2,300.[25] Some districts incurred much higher costs. The erection of cables and transformers to bring power to Glen Convinth, to serve 160 premises in an area of some thirty square miles, was costed between £60,000 and £70,000.[26] The cost of an 11 kV line, eight and a half miles long, to bring power to the district of Achavanich, in Caithness, was said in 1961 to be £1,000 per consumer.[27] H. R. F. Mackay, a resident of the tiny district on the long, lonely Causewaymire road across the rolling moor between the east coast and Thurso, had been trying for five years to persuade the Board to install a connection, while pointing out the irony that he was still lighting oil lamps when the pylons were crossing his land. The Board's defence was that isolated pockets were very expensive to connect but would be linked into the network as soon as higher Board income covered more of the annual losses.[28] But, as Tom Johnston never tired of saying, this was the reason the Hydro Board existed. It was the Board's first duty, he stated in February 1948, to guarantee supplies to the Highland population[29] and in each of its Annual Reports the Board's progress was boldly charted. A total of 116,000 consumers were connected in the six years until 1954. In April 1948, only 1,400 farms and 550 crofts in the Board area had had electricity, and now the totals were 6,273 farms and 8,566 crofts,[30] and demand was threatening to outstrip supply. In 1948 only about 4,000 houses in Caithness and Sutherland, almost all in the six towns and larger villages, had enjoyed a power supply; now, in 1954, 70 per cent of the population was supplied.[31]

Demand for electricity was growing in the Highlands and Islands at a rate above 10 per cent per year in the early 1950s.[32] In 1954, the Board calculated that it had spent almost £118.5 million on building power stations and distribution schemes.[33] By 1959, after fifteen years of Board activity, it was estimated that 90 per cent of the Highlands had been electrified and that one quarter of the region's water resources were involved in power generation.[34] The number of consumers passed the 400,000 mark in 1961[35] and 422,300 in 1964[36] as more remote areas were reached by the spreading network of power cables. The mains supply finally reached Barra in the summer of 1966 via an undersea cable from the diesel power station at Daliburgh on South Uist; 'It will certainly lighten our darkness,' quipped Father John MacCormick, chairman of the island's council, but, although the cable was duly installed, the islanders were still waiting for the power to be switched on over a year later.[37] Work began in the autumn of 1966 on the laying of 100 miles of main line and another twenty miles of low-voltage line to distribute the supply to North Uist and the islands of Grimsay, Baleshare and Berneray.[38]

When the Board celebrated its twenty-fifth birthday in August 1968, it could point to a list of achievements that boasted fifty-six major dams, fifty-four main power stations, almost 200 miles of tunnel, 400 miles of road either built or reconstructed, and over 20,000 miles of power line, all representing a total investment of £310 million.[39] It had done much since the first pole had been erected in Morar in May 1946. Electricity now reached some 96 per cent of the premises in the Board area.[40] The activity continued to expand – 10,000 new consumers were connected in 1970–71,[41] and the 108 houses in Kilchoan were finally served by an overhead line from Glenborrodale in 1972[42] – but by then the work of the Hydro Board was coming under serious scrutiny and its future was being closely questioned.

At the end of the Second World War, the national grid did not extend into the Highlands apart from one 132 kV line linking the Grampian company's Rannoch-Tummel scheme. There were several local distribution networks in towns and villages but these were isolated from each other. The Hydro-Electric Board, therefore, had as one of its primary tasks the construction of a distribution network to carry electricity to consumers scattered throughout the Highlands and to the national grid.

Dougie Maclennan started working for the Ross-shire Electric Company when he left school in 1935: 'We supplied up as far as Brora but the distri-

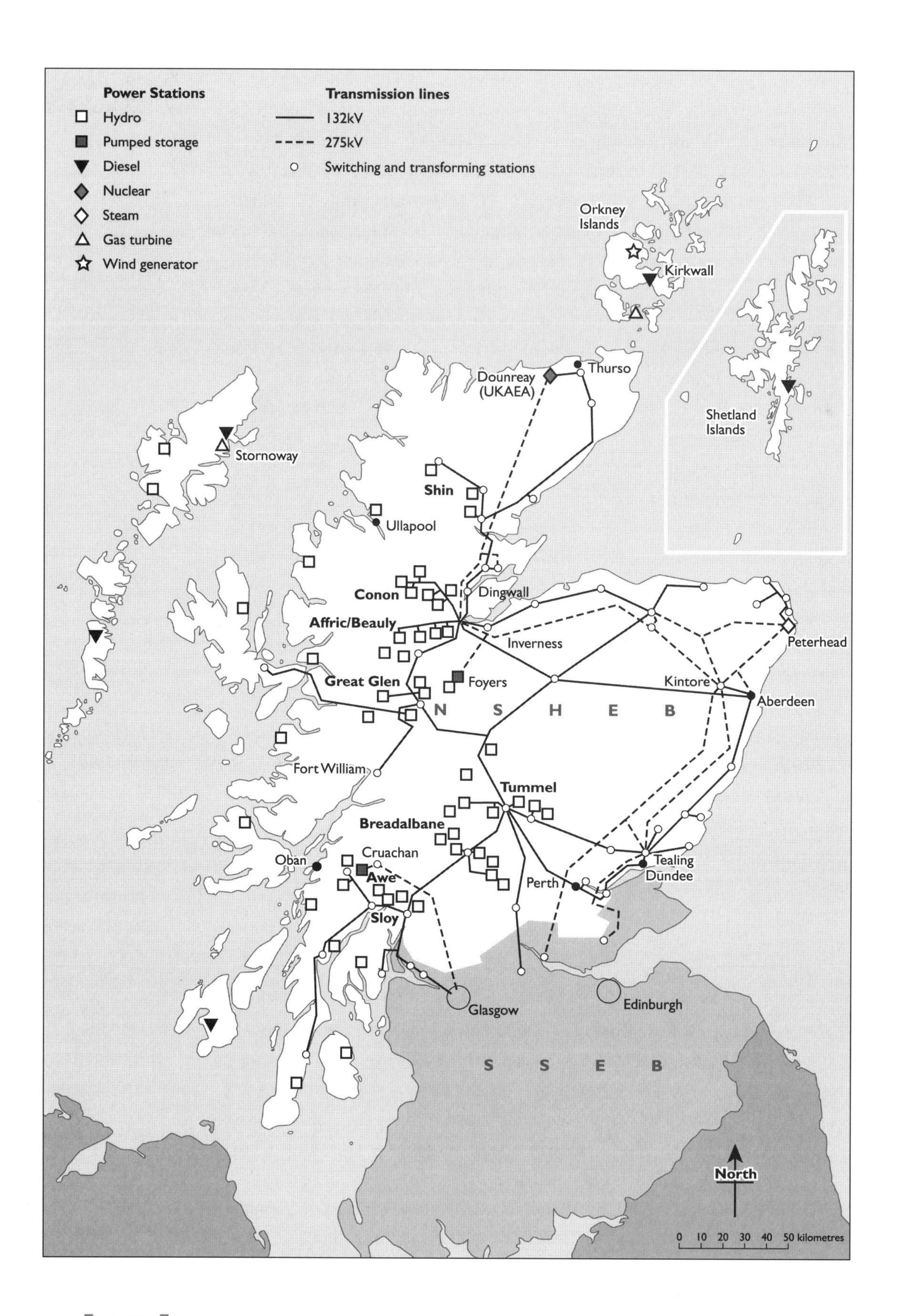

Power Stations
Hydro
Pumped storage
Diesel
Nuclear
Steam
Gas turbine
Wind generator
Transmission lines
132kV
275kV
Switching and transforming stations
Orkney Islands
Kirkwall
Shetland Islands
Thurso
Dounreay (UKAEA)
Stornoway
Shin
Ullapool
Conon
Dingwall
Affric/Beauly
Inverness
Peterhead
Great Glen
Foyers
Kintore
Aberdeen
N S H E B
Fort William
Tummel
Breadalbane
Cruachan
Oban
Awe
Tealing
Dundee
Perth
Sloy
Glasgow
Edinburgh
S S E B
North
0 10 20 30 40 50 kilometres

bution wasn't widely spread. Anything near the main line, which wandered all over the place, was on. The line went from Tain by undersea cable across the Dornoch Firth at Meikle Ferry, and on to Dornoch, Golspie and Brora. There was also a small diesel plant in Brora belonging to the Hunters tweed mill that ran at peak load times. Eastward we were sometimes supplying as far as Keith but I don't think we were supplying everything there – we used to swap readings on import-export [with the Grampian company]. Inverness had its own plant, and there used to be a little plant in Beauly village but it was taken over by the Ross-shire and, when the lines were extended, it was closed down. The Ross-shire also took over a small private plant in Evanton. The Americans had put in a 240-volt diesel power station at Invergordon during the First World War, and Tain had had its own wee plant attached to the town laundry until the Ross-shire took over and closed it down. The Grampian company took over the Ross-shire Company in 1938, and the Board took them all over in 1948.

Map 8.
Map showing power stations and transmission lines

'It was 1946 before I was discharged from the Army and I went straight back to the Ross-shire Company. Things were pretty ropey, after six years with virtually no maintenance. At this time people were installing cookers and so on, and we were getting overloads. My work was anything to do with the system. The Hydro-Electric Board was putting up new lines. The first power station was at Grudie Bridge and they took steel lines [pylons] down to Beauly substation. That was the first steel line. From Beauly it was eventually extended to Boat of Garten and into Inverness, and then of course on towards Elgin and further east. Then they took a steel line from Fasnakyle into Beauly and across the hill to Fort Augustus. Beauly was and still is a main switching station. Then Dounreay came in,[43] and the first thing they did was build a 132 kV line from Beauly to Dounreay. Before that there was no physical connection between the lines in Caithness and Ross-shire. Wick had its own diesel station. Later the Board built a 275 kV line that was tied up with the smelter project at Invergordon.'

The erection of the 80-95-foot pylons, the 'towers', for the high-voltage lines across the rugged landscape was a considerable achievement. The work was done by various contractors, including the J. L. Eve Construction Company Ltd and Balfour Beatty. Each pylon – they came in different shapes and sizes depending on the particular spot, and whether or not the lines were to alter in direction – was bolted together *in situ* and fixed to four concrete plinths. Jimmy Macdonald from Invergordon worked on the tower line down Glen Strathfarrar: 'We made a foundation, ten feet deep, for each leg of the pylon, but that depended on the side of the hill. We dug the holes by pick and shovel and sometimes we had to put in shuttering as we dug to stop the sides caving in. If it was rock, we drilled and blasted. I remember one fellow who put a box of geli' in for a blast and nearly shot down the helicopter going up the glen with the engineers. On the side of a hill it could be dashed awkward to get the gear in and build. Once the foundation was done, and the first section was up, you just built on and built on – you could do one [pylon] every couple of days. The bolts were left slack at first and were tightened when the whole tower was finished. The wiring gang came along then.'

The erection of the wooden poles for low-voltage lines and distribution within neighbourhoods was equally arduous and time-consuming. Once the Board's way-leave officers had obtained clearance for the route and once the distribution engineer had walked and pegged it and decided the appropriate height of pole, the erection gangs moved in. 'We took the poles and, especially in fields with crops, we carried them on our shoulders. Most of the poles were imported fir that had been pressure creosoted and, in summer with the heat, the creosote would have the face off you. We never got oilskins or boots or gloves.

'One fellow would put out the material for most of the line. Then a couple went ahead building. Another six would go poling. On a good day we would get eight or ten poles up. We used to walk them up on our shoulders, and put guyropes in to pull them, unless we had a fifty-footer when we had to use a flying jib. A rope to the head of the pole and a jib about five feet from the bottom, with a flying dolly on the top and a block and tackle. Pull that and, as the pole came upright, about to drop down in the hole, the jib and block and tackle would fly off. It was good but it was a big, heavy, iron thing. We only used it on hefty poles. You needed a high pole on the steep side of a hill to keep the wires off the bank as it went down. We packed earth and stones around the base of the pole with beaters, like ramrods – we called them priests. The only cemented poles were on one scheme where the rock was too bad for blasting.'

The work could bring out a flair for improvisation. 'Once we tried hitting a golf ball to take the line over a river but then we found a boat,' said Jimmy. 'Digging up on the hill we would keep the fags going to get some smoke around us to keep the midgies away. On the top of a hill like Ben More, you always took a billy can with you, to save going back for your tea. I've seen us dig a hole in the ground in the morning and later it would be full of lovely, clear water – for tea. Put a wee stick in it to take away the taste of smoke. We put creosoted wood in once – oh the taste. There was one place where we had been taking water from a burn for days, and then we found a dead sheep in it.'

'On a good day it wasn't bad work at all,' said Dougie Maclennan, a charge hand or foreman on the poling squads. 'The work was all spades and shovels. A gang would have three or four linesmen, and half a dozen labourers to dig the holes. To begin with, the poles were imported, properly seasoned and creosoted. In one year there was a big storm and a lot of forest was flattened down in Perthshire; it was decreed that these poles should be used, at least for spur lines. They were lying, unbarked, unsnedded, and they had all to be skinned and seasoned. They were all used but they gave a lot of bother later and had to be replaced, although for a while they brought power and light to some people.

'The hole to be dug depended on the height of the pole. An ordinary 32-foot pole would go in for five feet. Down in Easter Ross, where the digging was good, you could erect quite a few poles in a day. Things really took off after we started hiring JCBs; you could dig a hole in minutes. On the Black Isle it might take half a day to dig a hole with a spade. The distance between poles in a main line could vary, as we would be picking the ground a bit, but a long span might be followed by a short one, so that the two spans would be about average; and that depended on the height of the poles, the weight of the wires, and the ground clearance to allow room for sag in a hot summer. The wires are not insulated, except in certain sections, for example where the electric wires crossed above a telephone cable. In the early days the wire was copper, as thick as your finger on the bigger lines, some solid, some stranded, but now high power lines are steel-cored aluminium. It wasn't hard to handle.

'I saw a great deal of the country – anywhere north of the Caledonian Canal and Fort Augustus, the boundary of our district. We went as far north as Helmsdale, where we bordered the Wick area, and Durness and Cape Wrath; and we worked on South Uist for a while.'

Hamish Ross's father worked as a lineman for the Ross-shire Company in the 1930s, a job he combined with seafaring in the merchant navy when there was no work on land. 'He would just go to sea, to the Persian Gulf on tankers, and then work again for the Ross-shire after he came back, but eventually he stayed and worked on the lines until he retired in 1959,' said Hamish. 'When I started I was with the Grampian company for the first year and then the Hydro Board took it over. They started doing more lines and took on more men. I had joined after I came out of my national service in the Navy.

'The foreman in the Ross-shire Company then was Willie Macdonald from Strathpeffer, and there was a man called Ross from the Black Isle. He was a bit of a character: very strong, he could lift a 26-foot pole by himself.

Plate 88.
Dan Ross, Hamish Ross's father, using spikes to climb a pole during work for the Ross-shire Electric Supply Company in Alness in the 1930s (*Hamish Ross*).

They used to take the poles to the sites on two bikes, tied to the frames.

'I did a little on towers but not really very much – it was nearly all pole work. We used to work without gloves. Standing on the wooden pole, you didn't actually get a shock; you could work like a bird and connect live wires with your bare hands as long as you were on the pole and it was dry. You worked with only one wire at a time and as long as it was dry the current went only to your feet. If it was a wet pole and with your spike you'd get very, very little. The only thing you had to watch was that you didn't catch the connection and the wire at the same time. Then you became the jumper and that was pretty fatal. Now the safety is much stricter, with gloves and special eye glasses. On a wet day, you'd get a tingle in your hands but if was terribly wet it became much more difficult.'

The essential parts of the lineman's equipment were his spikes and his belt. The metal spike, about one and a half inches long, protruded from the side of the instep, in front of the heel, and enabled the wearer to climb and perch on a wooden pole. The spike and the supporting metal frame was fastened to the leg by leather straps around the knee and the ankle. A man adept on spikes could kick his way up a pole and stay there comfortably for a lengthy period of time while he worked on connections.

'When I came on first, they used to make them in the smiddy from old pram springs – good spring steel, and you needed steel that would bend a bit – you got exactly what you wanted to suit the length of your leg. You got to know your own spikes and after a while a pair belonging to somebody else wouldn't suit. Each person had his preferred angle of spike. My father was very good on spikes – everybody said that – he liked a straight one. I had them after he finished but I prefer them a little bent. They would get blunt and we would file them to sharpen them. Sometimes it might stick when you kicked it in, especially on old poles, but in the early days the imported Swedish or Norwegian fir was beautiful, like cushions.

'We had a leather belt with a brass buckle in the early days. To climb, you put your hand round the pole, threw the belt round and hooked it in, and you came to be able to do it without thinking about it. You might have to lie out [on the wire]. You might have to unhook and let the belt out two or three notches. I've seen men lying sideways in their belts for long enough, with one spike in and the other foot against the pole. Over the years men could develop neck pathology. I suppose the first few times when you started climbing you were a bit scared but you were frightened to say anything – it would look bad in front of the others. After a while games developed – see who could go up fastest. The worst I used to feel was if you were up and it was extremely cold with frost. Your foot would go kind of dead and, coming down, you'd find you'd lost the feeling and couldn't coordinate. When you reached the ground you staggered about for a second or two until the circulation got going.

'There were different sizes of cable, and some could be heavy to work with. On low voltage lines you had Number 6 wire. Number 8 was for the service to a house. Then there were Number 3, Number 1 and what they called 'Point 1'. That was about as thick as my thumb and I knew only one man who could bend it round an insulator.

'Many of the best linemen came from the east coast – Aberdeenshire and round that way – and they took a great pride in doing a good job. There was a place at Munlochy [on the Black Isle] where you could see for about three miles and you would think it was one pole, they were so well lined up, and they were all put in by "hand". With diggers now, the potholes are big and the lining up of the poles is not as good.'[44]

The final stage in connection was, of course, the linking of individual consumers to the network. This was usually taken care of by local electrical businesses who flourished in the wake of the Board's activities, wiring buildings for the first time or updating aged wiring originally installed by local suppliers. One such was Campbell Brothers in Dingwall. Ben Bentley, who settled in Ross-shire after he married, joined the firm at the end of the War: 'We took the cables into the houses from the nearest pole by whatever way seemed best. We passed porcelain tubes through under the flashing at the chimneyheads. As the firm grew, I grew within it and after three or four years I was in charge of men working as far down as Invergarry and through to Ullapool and the Black Isle. We didn't go so far north or do the remoter areas in the west.

'We had fifty-seven people outside on house-wiring at the peak, including labourers. At this stage, the houses were connected for free as long as they were within an area on the line of the main cables. If they were off the beaten track, they had to pay for a spur line. There were quite a lot in that position. The average cost to connect a house, for example a bungalow in the town, to the mains through the gable or the chimney would have been in those days £20. It wasn't cheap. That's what the Hydro Board paid the firm. We agreed a labour rate with the Board for actual work, time on the job and travelling, plus costs of materials and a percentage. The Board was a great outfit to work for, TJ was a great man, his spirit penetrated all the way down.

'We connected the houses as far as the position of the meter. The Board man came and put that in. Half the time it was our men who did the wiring inside the house. The light was the first to go in, then kettles and radios. Television was just coming in but we couldn't really get it here – I was the third person in Dingwall to have a TV. The best seller in those days was an Echo, it cost £30, a lot of money. Then there was a set called a Dynatron, a beautiful piece of work as a cabinet as well as a set. Having the Dynatron meant you were really with it.'

The use of electricity before the Board schemes came on line was sporadic and sometimes parsimonious. In the early 1940s, there were only five 100-watt streetlamps illuminating Maryburgh at night and only a minority of private houses enjoyed a supply. Villagers were therefore thrilled to have electric lights hanging in their rooms for the first time and a ready supply of convenient energy for household appliances. 'The people who had to adjust most were those who lived in the old black houses on the islands,' recalled Ben Bentley, 'but they adapted very quickly. You'd be in one day wiring the house and the next day they would have an electric cooker.'

On 16 April 1955, James Stuart, the local MP and Secretary of State, inaugurated the construction of a distribution scheme designed to bring power to 750 farms and some 2,000 other premises in Moray and mid-Banffshire. Stuart set in motion the raising of the first of the 10,000 wooden poles required to extend the network over the sprawling countryside. The overhead line in this scheme totalled 365 miles, with 350 transformers – typical of the amount of equipment needed for the Board's task throughout its fiefdom.[45]

By 1960, the Board had built 4,100 miles of high-voltage and 1,600 miles of low-voltage line at an average cost per consumer connected of £262.[46] The varied combinations of topography and population resulted in some distribution schemes costing far more than others: the expense ranged from Orkney at the low end of the scale (£120 per consumer) to Mull (£575) and Kames/Craignish and Cassley/Durness (each £660). The costs had one feature in common – they were almost all greater than the original estimates. The Board recognised, however, that this was why it had been created – to bring electricity to even the remotest places.

By 1960 it was estimated that a quarter of the Board's consumers were being supplied on an uneconomic basis and that the Board was losing a further £1.75 million in maintaining the service. The Board stuck to a tariff system that the majority of their customers could afford, although consumers did not always appreciate their good fortune.

'The Board brought a new prosperity and a benefit to the social wellbeing of the Highand communities,' said John Farquhar Munro MSP. 'When the Nostie Bridge power station was switched on, I was going to school in Plockton; I remember my grandmother, with whom I was staying, showing me the switch on the wall. There was a light in the kitchen and a plug; very few people thought of putting a light in the bedroom. It took a little time for the older generation to come to terms with the new easily obtained power – that you could plug in a kettle or an iron, that you could forget the need to conserve radio batteries, that you could buy a machine to do the washing.'

Stories began to circulate encapsulating the adjustments people had to make in their lives to accommodate the new source of power. No doubt a few are apocryphal, more of the nature of rural myth than actual event. One such might be the story about the old lady who thought the electric light was grand because she now had no trouble finding the matches for the Tilley lamp; but I have on excellent authority a true one about two elderly spinster sisters in one village. Three or four days after the spinsters' house was wired, the electricians received a postcard reading 'Dear Sirs, the electric light was wonderful when the man turned it on first of all but it has never come on again since he left the house'. The electrician went out to see what was wrong. 'Are you there?' he called as he went in. He put the light switch on and, of course, it worked. 'Well, well, isn't that wonderful?' said one of the sisters. 'Every man to his trade.' She went on: 'Do you know this? We've been sitting waiting every night. The streetlights would come on but not a thing would come on in here.' The old ladies had never realised the purpose of the switch.

A 'sparky' or electrician was somebody very special at that time in the rural areas but not all were to be trusted. In one west-coast district, one such man wired all the houses but went away to another job before the distribution lines brought power into the area. On the day of the switch-on, nothing happened in about 50 per cent of the homes. Then it was discovered that the electrician, who may have run out of money or equipment, had installed plugs and ceiling roses but no cable.

Some other folklore and expressions of popular humour have survived from this time. On the wall of an abandoned house near the Orrin dam I chanced on the following verse:

> Doon the glen came the Orrin men,
> they looked like ballet dancers;
> one in ten were time-served men,
> the rest were bloody chancers.

It was almost inevitable that the legendary figure, the Brahan Seer, should be credited with predicting at least one or two aspects of the electrification. The prophecy 'The day will come when North Uist will be encircled with steel' is held to refer to the building of the distribution lines; and 'A loch above Beauly will burst its banks and destroy in its run a village' was claimed to have been fulfilled when exceptionally heavy rain in 1967 caused the Torr Achilty dam to overflow and the flood caused considerable damage in Conon Bridge.[47] This was the prophecy to which Sir Edward MacColl made reference in his speech at the opening of the Affric scheme in May 1947, causing mirth when he added that the engineers had been instructed to use extra cement.[48]

During the years before the advent of remote control systems, power stations had to be manned in the night shift often by a single individual. 'There were long periods when it was very quiet,' said Ian Sim. 'You took a good book and you used to have a wander round the station every so often to check everything was okay. Occasionally on night shift at Inverawe you used to hear voices. You could never make out what they were saying but you knew they were voices. The first time it happened to me I thought there's somebody in here, and I looked all round the station and could see nobody. I even climbed up on the crane and went all round the outside. This went on for months. I ignored them, they never bothered me but you were aware of them. I didn't find out for a number of years that all my colleagues had also heard these voices on night shift but nobody had said a word. We never found out what caused it. There was a young chap who was sent down to Inverawe one time, and he came back and said to me, after he'd been down a couple of times, that he had a hell of a fright, he heard voices. I said join the club. He had heard them, stepped outside, and put his foot on this big frog which sort of exploded. He nearly had a heart attack, he didn't know what was happening. He fled inside, stood for five minutes to get his breath back, went out a side door with a torch, and saw the remains of the frog.'

Some of the stories reflect the dangers inherent in the work and contain motifs familiar in many folkloric contexts. One recorded by Cameron McNeish tells how a man was warned by a mysterious stranger not to go to work that day; the man heeded the advice and later learned that an accident in the tunnel where he would have been working claimed the lives of several of his friends.[49] The premonition of Margaret Chisholm in Glen Cannich, in this book's introduction, is matched with another from the same place concerning an old lady who said she could hear witches singing – where the pylons and the power lines now pass.

The men who built the schemes knew they were taking part in a great enterprise. In the early days of the Breadalbane scheme, Ronald Birse took time out to visit Loch Sloy, Strath Tummel and Pitlochry to see some of the already completed projects: 'We knew that hydro-electric power was contributing to Scotland's economy, and I felt certain it was a good thing – clean, renewable power.' Hugh McCorriston agreed that the weeks spent tunnelling through the rocks of Glen Affric had been worthwhile and that there had been a sense of achievement; when he tells visitors now about what was done, he said, they seem to treat him like a hero. After the construction phase, Hugh stayed on as a Hydro Board employee in Cannich where he, like many of the men who came to work on the schemes, met his wife and settled down.

'We never knew what to expect from one day to the next; we might be called out to any machine – at a moment's notice,' recalled Archie Chisholm. 'It gave me a broad knowledge – from compressors to pneumatic tools – it was a great training. I was so lucky to get five years of it, and so many experts around. In those days you couldn't just pick every part off the shelf, you had to make some of them. Of course the Board was strangled a wee bit by the government then because they were only allowed so much – if they stepped over, they couldn't get more. There was a credit squeeze on in 1951, you had to make do and mend.'

Archie went from working on the Cluanie scheme to do his National Service in the Royal Electrical and Mechanical Engineers: 'All the rest of the lads thought "what have we come to" when they saw the billets. I thought we were in paradise compared to the schemes, and I was well kitted out for this – a proper billet with proper facilities, with a big washhouse and everything spick and span.'

'The Hydro Board was a great thing,' said Ian Sim. 'It made life a lot better for many people. One great thing about it was that it was small – you knew people, in Edinburgh, Pitlochry and everywhere else, as you moved around. As engineers, we covered everything. If something went wrong, you fixed it. It was satisfying and we enjoyed it.'

By the time the work was finished, some of the men had been bitten by the roving life of the constructor and stayed in it until they retired. Patrick McBride remained thirty-eight years with Wimpey and, from Pitlochry, went on to help build, among other things, a reservoir in South Africa, a hydro-electric scheme in Turkey, underground oil tanks in Malta, a manmade island in Bahrein, power lines through the Saudi desert and a motorway in Nigeria, before retiring as a works manager.

He told me, though, how life did not always work out so well for others and gave as an example a Donegal man he had known at Pitlochry. This man was a loner and lived in a little cabin he had built for himself. Fond of a pint and a nip, a modest amount he called his quota for an evening, he also kept a bottle in a rubber boot. The other men, some of whom he liked, would always call on him and bring him milk or eggs if he needed them. He had a family at home in Donegal but seldom saw them and, after one visit when his youngest son was a year old, he left for good. Whether or not there had been a falling-out, no one else knew. Then, when this man and some workmates were checking in to a boarding house in Kilmarnock prior to a job in Ayrshire, he happened to bump into a young fellow who gave him a strange look. It turned out that the young fellow was his own son whom he had last seen in a pram in Donegal but who was now grown to manhood and a travelling workman like his father. They became great pals and travelled back to Pitlochry together.

As has been noted, there was considerable opposition to the schemes and great fear over the potential destruction of the scenic beauty of the Highlands. In June 1948, Mary Grant wrote from Kessock to the *Inverness Courier*:

> It is difficult to understand the complacency displayed by the press and public towards the monstrous schemes of the Hydro-Electric Board, and which, I presume, is only to be accounted for by the fact that we live in a grossly materialistic age. One cannot believe that at any past time the idea of putting 2,000 navvies to work in these wild and beautiful glens, destroying every familiar landmark, with all the legend and history attached, would have been tolerated. And yet, that is what is already happening in Glen Affric, where the work has been in progress for a considerable time. I am told that the conditions there are indescribable and the destruction of bird and animal life is complete.[50]

The *Courier* published almost no other letters in this vein; perhaps the editor, a staunch supporter of the Board, spiked them or, more likely, he did not receive them.

Seton Gordon, one of the most prominent naturalists of the time, was of the opinion that many a West Highland crofter would prefer a road or running water to electric light and that the schemes would not repopulate the glens. One of the most beautiful stretches of the Tummel had been inundated, with a loss of bird habitats, but even he had to admit that, after a spate of rain when excess water was being released, Loch Laggan dam was impressive and 'awe-inspiring'.[51] Frank Fraser Darling, a leading ecologist and advocate of Highland development, was strongly supportive of Tom Johnston's ambitions.[52]

W. H. 'Bill' Murray, one of the country's best-known climbers, was commissioned by the National Trust for Scotland to list the most outstandingly beautiful parts of the Highlands. Murray, who published his survey in 1962, set a high standard: of the western part of Glen Garry he wrote, 'Dams, pylons and the loss of woodland have shorn its old worth'. It was not all bad. He thought some of the newly created lochs, for example Glascarnoch on the Garve-Ullapool road, enlivened previously dull moorland, and Loch Faskally could be said to compensate for the damage wrought in the Tummel valley. Some of the roads, bridges and power stations had been well designed, he said, and blended with their locales but he expressed a particular dislike of the marching ranks of pylons.[53] This opinion was widely shared – full marks for everything, but a shame about the pylons, to put it crudely.

Perhaps it just needed time. The authors of *The New Shell Guide to Scotland*, first issued in 1965 and finally revised in 1977, included a whole appendix devoted to the schemes. In 1969, W. Douglas Simpson, while agreeing that Pitlochry had 'suffered somewhat', found inspiration in the same power lines: '... at more than one point along this Great North Road [the A9] we are accompanied by the pylons, striding along the skyline – in their fitness for their purpose asserting a just claim for recognition ... as no unseemly contributions to the ever-changing Highland landscape. How proudly do they pursue their way across hill and brae and howe and glen, proclaiming their beneficent mission of bringing the comforts of modern life to remote clachan and lonely cottage, and, thereby, playing their part in arresting the canker of depopulation in the Highlands!' [his exclamation point].[54]

Simpson's enthusiastic acceptance of the pylons seems not to have been read by Jimmy Page, the lead guitarist with the rock band Led Zeppelin, who instigated a protest in 1971 against the Board's plan to push a power line along the south-east shore of Loch Ness from the new scheme at Foyers past Boleskine House, the mansion once owned by the infamous master of the black arts, Aleister Crowley, that Page had recently bought. The musician wanted the cables to go underground but the Board argued that this would cost too much and would make fault location difficult. Page organised a protest outside the Board offices in Inverness and a public meeting, and collected a thousand signatures on a petition.[55] At a public inquiry into the problem, the Board explained that burying the cable where it crossed Page's

Plate 89.
The dam site at Cluanie before construction started (*Mitchell Report No. 1. Ronald Birse*).

Plate 90.
Pylons and power cables (*author*).

Plate 91.
Once a river valley and now a loch; the upstream view from the Glascarnoch dam in May 1957 (*NOSHEB*).

land would cost an extra £55,000, that putting it underground all the way to Inverness would set them back a whopping £450,000, and that submerging it in Loch Ness would cost even more. An overhead line was estimated to cost £40,000 per mile.[56] The Secretary of State, Gordon Campbell, finally allowed the overhead line but ordered that the section past Boleskine House should be buried.[57]

As if in a late riposte to Mary Grant of Kessock, the *Inverness Courier* quoted in 1972 a letter by A. K. Brandon, London, to the *Daily Telegraph*. It was only to state, said Mr Brandon, that the Board had not destroyed the natural beauty of Glen Affric but had 'added to its attractions and ... left no noticeable reminders of how their difficult task was accomplished'.[58] In fact, the Board had gone to considerable lengths to avoid disrupting the scenic quality of Glen Affric.

Elsewhere, some significant changes were wrought on the landscape. Ronald Birse recalled the creation of new lochs: 'There's a big loch now up Glen Lednock above Comrie, where the main dam is. The loch there is four to five miles long. I don't think it actually submerged any habitation. [People having to move home] was relatively rare, and we didn't need to cause that, fortunately.' Relatively rare though they were, enforced flittings did take place. 'The railway station at Luichart went underwater, as well as a number of houses, when Luichart became twice the size it had been,' said Donald Macleod.

The coming of the dam builders suddenly brought a swirl of activity to places where life had been running down for years. John Farquhar Munro saw the changes in the Cluanie Bridge area. In his boyhood the only road was the single-tracked, narrow, winding A87 through Glenmoriston. The Cluanie Inn had closed down in 1940 and the nearest telephones were fourteen miles away down the trough of Glen Shiel and twelve miles south over the hills at Tomdoun by Loch Garry. There were only four pupils – John and three brothers from another family – in the local school, a Side School, located in a bothy attached to the shepherd's house at Corrie More and later transferred to the parlour of the former Inn. For a time, the teacher cycled every day from near Dornie, a round trip of forty-odd miles.

'You can imagine that in such a remote place we were amazed at the amount of sophisticated plant and the scale of the operation [when the scheme started]. We had no power of any sort, not even a private generator. We had the Tilley lamp, the candle and the Primus stove,' recalled John Farquhar Munro. 'The Cluanie Inn started up again as a hotel and did good business. We were seeing things we'd only seen in books, things like excavators, diggers, big lorries, cranes. Another innovation was a blondin across the top of the dam. I suppose the most startling was the concentration of people in one area. There was a big opportunity for the people down in Glen Shiel, Dornie and Kyle – I used to hear many of them years afterwards talking of their experiences at the dam. Crofters who became concrete inspectors and got very good jobs, not because they had any qualifications but because they understood what was going on and had a good sense of right and wrong. Many of them went on from the Cluanie dam, continuing this sort of discipline, right through the era of the hydro schemes.'

The reaction of John Farquhar Munro's father to the building of the Cluanie dam and the creation of the eight-mile-long Loch Cluanie between Glenmoriston and Glen Shiel was more mixed: 'Before the dam was built, there was at the east end [of the glen] what they called the Big Loch, two mile long and maybe half a mile wide, and a narrow cut and then the Wee Loch, maybe 200 yards across. I don't think there were many objections to the scheme. My father took great exception to the desecration of the valley but I don't think he was looking at it from an environmental point of view but from his own situation – he lost his arable fields, he lost his hayfields, all the fencing went underwater. Worse still, the place was flooded about September – I can't tell what year, say, 1955 – when he was in the middle of securing the hay. He had three-quarters of it ready to be harvested when the water level rose. I remember him making rude noises to some of the senior people. They gave him the equivalent of the tons of hay lost, but that was only for the one year. That was aggravating for him, and he didn't get any benefit from the scheme. There was no electricity put into the lodge, the hotel or any of the residences near the loch. All the power went off the other way, over to Glen Garry and Fort Augustus. The little community most adversely affected didn't have any benefit. The electricity never came and the locals are all on diesel generators to this day.'

Archie Chisholm's family had already left Glen Cannich by the time the hydro-electric scheme came to be built. Archie's grandmother, who had sensed the work to come, never went back: she said it had been spoilt for her and she didn't want to see the one large loch now formed where she had known two smaller ones.

Iain Thomson and his family were the last people to live in Strathmore at the west end of Loch Monar before the loch was dammed and enlarged in the Strathfarrar scheme. The hard, yet in some ways idyllic, life of the hill shepherd is well described in his book *Isolation Shepherd*, which also gives a glimpse of what was lost when the schemes were built. There must have been many a glen dweller like Thomson's neighbour Kenny Mackay who, reluctantly accepting 'progress' as his world changes irrevocably, said: 'Mind you, boy, the light in the byre is handy but all the power in the world couldn't replace the pleasure of a fine day on Loch Monar.'[59]

[6]

'... still a huge part to play ...'

Once a hydro-electric scheme was built, a team of Board engineers and technicians took over the running of the turbines. Ian Sim was one of them: 'It was 1956 when I started with the Hydro Board and I was with them for the rest of my working life. We stayed in a converted hut that was originally beside the work camp for the Errochty scheme. They split four of the huts in two and made them into semi-detached houses, nice wee houses, set on a brick base with a nice garden and a nice outlook. It was very remote in those days, with just a narrow twisting road down to Pitlochry, that ran up to Rannoch Station in the other direction. There was another road over the foothills of Schiehallion to Aberfeldy. There was a locked footbridge across the Tummel in front of the house, for cables, and we had a key and could go across this bridge. It was only five minutes' walk from the control centre.

'The whole Tummel scheme was controlled from there. There were engineers at Rannoch power station who ran it and a very small power station at Gaur, under our instructions; and engineers at Clunie who ran it and the Pitlochry power station. We got our instructions from the control centre at Portnacraig House in Pitlochry. Each hydro generation group, the power stations in a catchment area, was treated in effect as a single station. They would phone up and say we want 100 megawatts or 150, and you then distributed that load amongst your power stations as you needed to, taking into consideration the levels in the various reservoirs, rainfall, and run-off. They wanted the power supply, as they say, yesterday and the period of time could be forecast fairly closely. They worked from records and looked for the demand for the same day a year ago, adding an additional figure based on the increased load in the area. From the records, they would know that the load in the morning built up until half past eight, and then stayed even or pretty steady up until 12 o'clock when people stopped working and went for lunch, and then built up again in the afternoon to the tea-time peak and then dropped off again. At that time, their boss, the man in charge of the whole grid for the Board, would phone up and speak to our boss, the generation engineer for the Tummel valley, and they would discuss what should be done. We had a good idea how much power we could generate from each power station in the coming week, and we would arrange a programme, splitting the load between each power station and working out what this meant for every reservoir. The boss might say, for example, he didn't want Loch Rannoch drawn down by six inches, and we would adjust the programme until we got something that suited.

'You could relate load to the size of the reservoir although the one imponderable that hung over your head like a black cloud was what was the weather going to be. We had one or two rain gauges and at Tummel itself the Board had an agreement with the Met Office whereby they installed the equipment and we read it every six hours night and day. This caused a bit of stramash with the Met Office. The wet bulb of the wet and dry bulb thermometers was in a small glass bottle and in winter they froze and cracked. We had terrible frosts at Tummel Bridge. When the thing thawed the water ran out. We would phone up to say we needed another glass bottle, and this chap came all the way up from Bracknell or somewhere to show us how to deal with the glass bottles because they had never had a place that used so many of them. We told him we should go to plastic and eventually they did this. They admitted they had no idea that temperatures in the Central Highlands could go as low as we experienced. During the summer they had no idea it could get so warm.

'The thing you had to try to avoid – it was a black mark against you if you didn't – was if you had to spill water from the reservoirs. Too much water, and that water had to be spilled from the dam down the river. That was waste, no use to man or beast. It was inevitable that this was going to happen and it used to occur every year in the autumn and the spring. A lot of the spring flood was due to snow melt; in the autumn it was just due to the rain. What you tried to do was have enough room in your reservoirs for the autumn so that you could absorb a lot of the rain; you could never absorb it all. The trouble with our reservoirs was that they could never be built big enough to give seasonal storage, so that you could store all the rain that fell over the autumn season and the winter. You could never get permission to build a dam big enough to do that. It's a question of economics – dams and tunnels are expensive to build, and you have to accept building at an economic level and the loss of water at various times of the year.

'The Board had a series of agreements with landowners, and what they had in effect was a maximum and minimum operating gauge. The maximum was set by the spillways in the dams; the minimum was set by these agreements. The Board had to pass compensation water from the dams to maintain river flow and very often in Perthshire and Argyll this was because of fishing. Salmon fishing was very important. They had a strict compensation programme at Pitlochry. Because people waded into the Tay up to their chests to fish, you could not vary the water level during daylight hours. The power station had to be set by early morning and it had to stay at that level of flow until the following night. Obviously, if you were in a flood condition, there was nothing you could do but, for normal operation, the compensation was very important. I remember many years ago the then chief executive of the Board (Angus Fulton) was visiting and he was having an argument about compensation flows. He was adamant: whatever happens, compensation flow takes the top priority. During the fishing season, on the Tay from 5 January right through, what we did in working out the programme was to draft a rough programme based on the output of Pitlochry power station for the months ahead. Then every week we had to draw up a very precise programme for Pitlochry and variations had to be taken into account at other stations, which meant that you had to have sufficient room in the reservoirs to allow for daily variations. You could not vary Pitlochry power station. At that time we did the planning by hand – you sat down with a piece of paper and a pencil, and worked it out. You could relate kilowatt output directly to cubic feet of water.'

Right from the beginning, the Board was conscious of the need to cope with the importance of salmon fishing in the Highland rivers. The 1943 Act that brought the Board into being made regard for fish stocks a requirement and the Fisheries Committee, with Colonel Sir D. W. Cameron of Lochiel in the chair, was established from the start especially to advise on this issue. The plans for each scheme included provision for compensation water, as the amount of flow left in a river system to maintain the fish population was termed. Fishing owners and others were to claim that the fishing in some waters deteriorated after the Board had constructed a scheme on a river system but, in the absence of good 'before and after' data, these claims were to be rarely upheld.[1]

In some places, the schemes increased the range of the salmon. Before the construction of the Loch Luichart dam, the Falls of Conon had proved an impassable barrier to the fish but the installation of a fish lift allowed salmon access to the loch and the upper stretches of the river Bran for the first time.[2] A fish lift or fish pass was included in the plans for many of the dams, perhaps the most common design being the one named after its inventor, Joseph Borland. In this device, the migrating salmon are attracted by the

Plate 92.
The Orrin dam under construction, showing in the centre two of the four giant pipes comprising the fish pass, October 1958 (*NOSHEB*).

water flow into a chamber at the level of the tailrace, a sluice gate is then closed, and the water is allowed to rise in a tunnel until it reaches the level of the upstream reservoir, carrying the fish up with it. The Clunie and Pitlochry dams were fitted with long fish passes or ladders, essentially a series of pools connected by short tunnels through which the fish can swim past the dam. Smolts migrating downstream can, amazingly enough, pass through turbines unscathed but in the larger schemes screens were placed to guide the young fish to the safer fish pass.

The Board also went to considerable lengths to ensure that spawning beds were adequate, or it undertook stocking. The new Loch Faskally at Pitlochry was stocked with 6,700 yearling trout in May 1951, with Tom Johnston, himself a keen angler, enjoying the chance to pour in the first batch. It was reported that the artificial loch looked natural and that, far from damaging the town's reputation as a resort, it was now being promoted as one of its attractions. The Board had also seen fit to erect a cedarwood boat house with twenty boats for the use of anglers and trippers.[3]

It became normal practice for the Board to buy the fishing rights on a river and, when the scheme was constructed, hand over supervision of the angling to a local association. In June 1966, when the Board let it be known that it intended to sell the fishing rights on the Conon and the Blackwater back to the private owners, there was a great outcry.[4]

To compensate for the loss of salmon redds in the river Garry, the Board established at Invergarry what was then the biggest hatchery in Europe, able to rear seven million eggs. The adult fish were caught in the late autumn at a barrier across Loch Poulary, to be stripped of eggs and milt, and the fry were released in the following spring. A Borland fish pass was installed beside the Garry dam.[5] By 1959, the Board was operating a third hatchery, at Contin, in association with the Conon District Fishery Board, where two million fry were raised for release into the Bran. Other measures to protect salmon stocks included a programme of killing pike in Loch Luichart and other lochs.[6]

Fishing interests were considered in the public inquiry in Edinburgh in December 1957 into the Strathfarrar scheme. The Fisheries Committee failed to persuade the Board to alter the plans of the scheme and had to accept guarantees on compensation water for the Farrar and Beauly rivers. William J. M. Menzies, the Board's fishery adviser, told the inquiry that with the proper use of compensation water he saw no reason to fear an adverse effect on angling. Other witnesses testified that angling had improved on some stretches of the Blackwater after the Board's operations.[7] The final inquiry report to the Secretary of State concluded that only 17 per cent of spawning

20/10/58

Plate 93.
The completed Orrin dam, November 1960. The fish pass is in the sloping concrete structure in the centre (*NOSHEB*).

grounds would be lost and that fish stocks as a whole would 'not be seriously diminished'. The possible damage to angling on the Beauly, it said, was not enough to hold up the hydro scheme.[8]

The fears that the hydro schemes would desecrate the landscape and bring disaster on the Highlands' tourist industry have never been realised. Few could claim that a dam such as Glascarnoch right beside the main road between Dingwall and Ullapool is an object of beauty, but it is just about the only structure of its kind. Other large dams, such as the Mullardoch, are tucked away in remote glens and are probably seen by only a few hardy hillwalkers and local people.

The native Highlanders raised few objections to the schemes on amenity grounds. Councillor George Grant of Golspie was quoted in June 1950 as saying that his county (Sutherland) 'wants the song of the workman and the milkmaid rather than the raucous screech of the eagle'.[9] For a time, the schemes brought employment and money into the Highlands and soon began

Plate 94.
Iain Macmaster (on the left) and some workmates at Strathfarrar in 1964 (*Iain Macmaster*).

to attract tourists. Welcomed though it generally was in the Highlands, the Board found itself the target of serious opposition in other quarters and had to fight hard at times to protect its independence and its mission. As the early decades of the Board's existence passed, this opposition grew bolder and, with shifts in the economic climate and social values, began to sniff victory.

Aims of Industry, a right-wing organisation opposed to nationalisation, was a strong foe. Sir Gerald Nabarro, a flamboyant Conservative MP who often had a go at the Board, called the Loch Sloy scheme a white elephant in 1955[10] and often returned to the attack. A House of Commons Select Committee on Nationalised Industries praised the Board in December 1957 and rejected complaints, for example by Sir David Robertson, the MP for Caithness and Sutherland, that remote parts of the Highlands were not being supplied fast enough. Sir David Robertson had already made allegations in the House that the Board was exporting more electricity to the Lowlands than it was supplying to the Highlands; this charge was repeatedly refuted by Tom Johnston and other Board staff.[11] Sir Gerald Nabarro proposed a Commons motion to reject the Strathfarrar – Kilmorack scheme in July 1958 but mustered only two votes (his own and Sir David Robertson's) when the issue was put to the test in a late-night parliamentary sitting.[12] His argument that the English taxpayer was being asked to pay for 'the unwanted extravagance in the far north' cut no ice, and neither did Sir David's opinion that the Board was reckless in pursuing one expensive project after another, a policy he termed 'squandermania'.

The Highland press was almost unanimously in favour of the Board's ambitions, with the *Inverness Courier* and its doughty editor, Dr Evan Barron, leading the home side. This allegiance did not prevent Barron from occasionally reminding the Board of its duty, but he was more often fierce in its defence. When the Local Authorities Committee griped about the Board's tariffs in February 1956, the *Courier*'s leader attacked it without mercy: 'a crasser exhibition of ignorance, hypocrisy, prejudice, stupidity and childishness has seldom been witnessed in the Highlands'. As if winded by this jab, a few months later the Committee was considering whether or not it had outlived its usefulness.[13]

In October 1956, the Scottish Landowners Federation argued that more schemes were not justified and called for a government inquiry into the whole question of hydro-electric development.[14] Six months later, the Board published its plans for the Strathfarrar-Kilmorack scheme. The project was estimated to cost £14.25 million and would generate 261,000 kW of power. It would take five years to complete; the principal elements were dams on Lochs Monar and Beannacharan to feed underground power stations at Deanie and Culligran in Glen Strathfarrar, and two dams at Aigas and Kilmorack on the lower stretches of the Beauly river. One thousand workers would be employed in their building. A more ambitious scheme, with the overall title of the Awe Project, was published in July 1957. Two sections of this scheme, a barrage and power station at Inverawe and a dam on Loch Nant to feed a power station near Inverinan, were conventional, but the third section was a first for Scotland – the construction of a reversible pumped storage scheme on Ben Cruachan itself. A dam would be built in a corrie high on the summit plateau of the ben as the reservoir for turbines in an excavated cavern deep inside the mountain. The water would discharge into Loch Awe but, during times of low demand for electricity, the same turbines would adopt the role of pumps and draw water from Loch Awe back up to the reservoir. Pumped storage had been a dream of Sir Edward MacColl's and now it was to be realised. The Ben Cruachan section would produce 400,000 kW, and the construction of the three sections, capable of generating 450,000 kW in all, was estimated to cost £24.5 million.[15]

The Secretary of State, J. S. Maclay, ordered a public inquiry into the

Plate 95.
Working on the Beauly river at Kilmorack, 23 October 1959 (*Press and Journal*).

Strathfarrar-Kilmorack scheme in November and the proceedings opened in the Signet Library in Parliament House in Edinburgh on Monday 9 December, R. S. Johnson QC presiding. Four objections to the scheme had been submitted. C. J. D. Shaw QC, Dean of the Faculty of Advocates, spoke for the Board and said the question to be decided was whether or not the scheme was in the public interest. He went on to submit that it was, because it was in the public interest to develop hydro-electric resources and distribute power to remote areas, and because the scheme would conserve coal resources. Appearing later as a witness for the Board, Tom Johnston stated the scheme would save the country 145,000 tons of coal per year. The threats to scenery and fish stocks were admitted but Shaw argued that it had to be accepted that the Board had to build its schemes in areas of high scenic beauty. The Board was conscious of its responsibility and had rejected cheaper plans for this scheme that would have meant flooding a 'fair amount' of arable land. With regard to the fishery, Shaw maintained that the proposed amount of compensation water was 'not only adequate but generous'.

On the second day of the inquiry, witnesses put forward evidence that the scheme would not adversely affect the fishing and some, with experience of other rivers, testified that fishing had improved after a scheme had been built. Another witness, Robert Johnston, an estate factor, thought no serious damage would result to agriculture. The proposed dam on Loch Monar would flood about 1,000 acres belonging to Sir John Stirling of Fairburn and 650 acres belonging to Lord Lovat; of this total, about 150 were potentially arable, said Johnston.

The first witness for the objectors was Dr T. Elder Dickson, vice-principal of the Edinburgh College of Art and a past president of the Scottish Society of Artists. He did not want another hydro-electric scheme in 'the last and loveliest of the glens untouched', and argued that the natural heritage should not be despoiled. Sir Robert Spencer Nairn, owner of the Struy estate, also argued in favour of preserving the unique beauty of the glen. On the third day, the third objector, John Mackay from Patt, Struy, presented a petition signed by the seven residents of Glen Strathfarrar. They protested against the depopulation of the glen, the flooding of winter grazing and arable land, and their having to move to seek an alternative living against their will, as well as against the damage to the scenery. They felt that further generation of electricity by such schemes was unnecessary and uneconomic. John Mackay worked with his father and brother as shepherds and stalkers on Sir John Stirling's estate; the prospect of flooding the low land around Loch Monar would deprive them of valuable grazing for their cows and sheep.

The inquiry lasted five days. R. S. Johnston visited Glen Strathfarrar in January to see the area for himself. Perhaps as might have been expected, his report to the Secretary of State in February accepted the Board's evidence that the increasing demand for electricity made the works essential and that they were ultimately in the public interest. 'This is not to say that it is not in the public interest to preserve the natural scenery or to leave uninjured the stock of salmon in any river, both of which may also be regarded as national assets,' wrote Mr Johnston. 'A balance must, however, be struck between the competing claims of electricity on the one hand and the preservation of the natural beauty of Scotland and salmon fishing on the other.' He had 'considerable sympathy' for Mr Mackay but pointed out that his employment by Sir John Stirling was not in question. On 5 June 1958, the Secretary of State issued the Order to allow the scheme to proceed.[16]

The arguments during the inquiry about the economics of the scheme had mentioned the prospect of nuclear power making hydro-electric schemes redundant in perhaps ten years' time. The United Kingdom Atomic Energy Authority had already started developing the nuclear power station at Dounreay on the north Caithness coast with the dream of producing cheap, limitless electricity from its reactors, and how to reach an accommodation with this new competitor exercised the Board for several years. In April 1970, the Board announced its decision to build its own nuclear power station at Stake Head, Banffshire.[17] Tenders for its construction were invited in January 1971 but before the year was out the Board was reconsidering the plan. Thomas Fraser, the chairman, announced in a press conference in the Royal Stuart Motor Hotel in Inverness that the £100 million scheme was in abeyance while the Government examined its policy for the nuclear industry.[18] Two years later, financial restrictions arising from the Government's anti-inflation measures led to the Board shelving any ambitions to develop its nuclear power station.[19]

There were questions in Parliament about the sums the Board was paying out as compensation to landlords whose property was affected by flooding. For example, it was stated that £200,000 had been handed over to Lord Lovat.[20] The Board was experiencing some financial problems in the late 1950s, and Westminster published a bill to increase the Board's borrowing powers from £100 million to £300 million in December 1958. The Board's Annual Report for 1958 admitted a loss of £112,238, a deficit that was ascribed to the rainfall having been 14 per cent down and the resulting need to buy compensating electricity from outside the Highlands. At the opening of the Glenmoriston scheme in February 1959, Sir Christopher Hinton, a leading expert on energy and industry and until recently the chairman of the Central Electricity Generating Board, addressed the issue: 'It seems to me that twenty, thirty and fifty years hence, people ... will not think of the temporary financial difficulties ... but will say how tremendously fortunate it was that this water power development took place when it did.'[21]

Hydro-electric schemes are vulnerable to the vagaries of climate change and summer droughts had also caused problems in 1955 and were to do so again in 1960, when the river flow was only 81 per cent of the expected average [22] and in the unprecedented drought of 1968-69 when rainfall fell to 71 per cent of its usual level.[23] The need to import power slowed the rate at which the Board, unsubsidised from the public purse, could clear accumulated debts. A wet year bucked up the accounts. The heavier rainfall in 1961, for example, reduced the need for power from thermal power stations, allowed the sale of 780 million units to the South of Scotland Electricity Board and secured a credit balance of over £256,000.[24] The wet winter of 1966-67 again prevented a loss.[25]

Sir John Stirling of Fairburn, convener of Ross and Cromarty and whose family owned large tracts of land, including the upper stretches of Glen Strathfarrar where the Monar dam was to be built, pressed the switch to close the last gate in the Orrin dam on 4 April 1959. The Orrin river, now thwarted on its seaward journey, slowly began to swell to form the new Loch Orrin, eventually to be five miles long. The construction had taken only three years and had been carried out by Duncan Logan Ltd of Muir of Ord. The gate-closing ceremony took place before a crowd of about 100 that included the Ross and Cromarty MP, Captain John Macleod, and the provost of Dingwall, Alex Macrae, who was also a member of the Board. Lord Lovat and Sir Robert Spencer Nairn sent apologies for their absence.

Barely two weeks later, the retirement of Tom Johnston was announced. The canny rabbit who had bargained with Churchill away back in 1941 had now reached the age of seventy-seven. His successor as chairman of the Board was named as Lord Strathclyde.[26] Thomas Galbraith, Lord Strathclyde, already had a career in the Royal Navy behind him, a long stint in Parliament as the MP for Pollok, work as an accountant, a period as a Glasgow councillor and four years as a Minister of State in the Scottish Office. 'He was an amazing man,' recalled Hamish Mackinven, 'a four-ring naval commander, and by profession a chartered accountant. I had been travelling with Tom Johnston when he would see a group of linesmen in a field, tell the driver to stop the car, and climb over a fence to introduce himself to them. The men would be amazed that this was the chairman come to speak to them. Now Strathclyde had the same ability to speak to the "lower deck".'

The Board lost its deputy chairman, Sir Hugh Mackenzie, on Christmas Eve that same year. Twice the provost of his native Inverness and knighted in 1945, Sir Hugh was a 'genial, approachable and friendly' man, and he had been deputy chairman of the Board since 1951.[27]

In November 1960, a group of industrialists, sportsmen and landowners formed the Scottish Power Investigation Committee with the specific aim of bringing about a re-appraisal of the Board's activities. The members of SPIC (author's acronym) were Sir John Craik Henderson, Michael Baillie, Major R. Macdonald Buchanan and Colonel W. H. Whitbread. 'When one looks at all the factors, they strongly suggest that thermal stations, generating in the south and distributing to the Highlands, would mean cheaper electricity and be better for Scotland,' Sir John was quoted as saying.[28] Michael Baillie, then representing Dochfour on Inverness County Council, proposed that the Council institute an inquiry into Hydro Board affairs – he thought the Board wasted resources – but the Council rejected this.[29] Colonel Whitbread was a member of Aims of Industry and was also anti-Board because its proposed scheme for Lochs Fada and Fionn, north of Loch Maree, threatened his estate. SPIC also picked up on publications by Denys Munby, an economist who had submitted evidence to the 1957 Select Committee on Nationalised Industries purporting to show how uneconomic the Board's hydro-electric developments were.[30]

Opposition to the Board's plans for Glen Nevis united Scottish MPs in its defence. Ian MacArthur, MP for Perth and East Perthshire, made a strong pro-Board speech in a House of Commons debate in December 1960: he accepted that the cost per kilowatt generated by a hydro-electric station was higher than for a thermal station but showed that this was offset by the fact that a thermal station had a life of only twenty-five years whereas a dam and power station had a much longer lifetime with low running costs. Regarding the amenity argument, he described how, on a walk through Glen Affric, he '... saw this great dam rising out of the glen [and was] astonished not by the harm it had done to the beauty of the glen but by the way in which it had somehow improved the aspect before my eyes ... a symbol of strength ... new life ... new hope'.[31]

SPIC was attacked for being self-styled and biased against the Board, but in March 1961 the Secretary of State, John Maclay, set up a formidable potential enemy in the shape of a departmental committee to review the whole picture of electricity generation in Scotland. The *Inverness Courier* objected to the Secretary's initiative, pointing out that, apart from James Shaw Grant, the editor of the *Stornoway Gazette*, no one from the Highlands and Islands had been appointed to serve on it.[32] At the same time, other views were being made public. An editorial in the November 1960 issue of *Scottish Field* was very critical of the Board's proposal to build a dam in Glen Nevis.

The departmental committee, known as the Mackenzie Committee after its chairman Colin Mackenzie, met for the first time in St Andrew's House on 6 April 1961 and continued to consider its remit during the rest of that year. Meanwhile, all the proposed hydro schemes were put on hold. Lord Strathclyde expressed the fear in December that the break in the continuity of the work would lead to a dispersal of the professionals, equipment and expertise acquired in the last fifteen years.[33] The Electricity Consultative Council had backed the Board and hoped that nothing would be done to endanger its independent status, and all the local authorities and newspaper editors in the north had taken the same view.[34]

Hamish Mackinven regarded the period of the Mackenzie Committee as a time of extreme danger to the Board. He knew that Aims of Industry had only one purpose – to destroy nationalisation – and was spurred into action one day when a letter from them arrived on his desk: 'I immediately thought why are they writing to us? A quite innocuous enquiry wanting to know something. But it was the name. I went to Angus Fulton the general manager. Fulton was one of the world's great hydro-electric engineers. If you had gone to the Soviet Union or the USA and asked for the ten finest hydro-electric engineers in the world, his name would have been included – he was world famous. But he was a total innocent as far as politics was concerned. I told him about the Aims of Industry and he wouldn't believe me. But I knew something was afoot.'

When the Mackenzie Committee began its work, Mackinven started to hold unofficial, secret press conferences in his Edinburgh flat to feed to the papers what the Board was doing. The official line in the Board headquarters was that the Secretary of State 'must not be embarrassed' but Mackinven continued his lonely campaign to defend the organisation he loved. The Mackenzie Committee finished its deliberations and published its long-awaited report in November 1962. As had been anticipated, it recommended the merging of the North of Scotland Hydro-Electric Board with its southern neighbour, the South of Scotland Electricity Board, to form one body covering the whole country. The editor of the *Highland News* called this 'an obnoxious proposal'[35] and there appear to have been few north of the Highland Line who disagreed with him.

By this time, Michael Noble had replaced Maclay as Secretary of State but the new man accepted the Mackenzie Report. Hamish Mackinven suggested to Douglas Neillands, secretary of the Board's Consultative Council, that the Council should call a meeting. The Council was made up of thirty-one members from throughout the Highlands and Islands and the chairman was Lord Macdonald of Sleat. Such a meeting was held in Perth and was attended by representatives from the eighteen local authorities in the Board area. It resolved unanimously to seek a meeting with Noble to impress upon him their opposition to the Mackenzie proposals.[36] The Mackenzie Report had placed a cash value of £300,000 on the Board's social clause but the Consultative Council argued before Noble that this was a gross under-estimation of its true worth.[37] The Mackenzie Report coincided roughly with the time when the Beeching cuts were being made in the railway network; Beeching had recommended the closing of all the rail lines in the northern Highlands, prompting Evan Barron to write, 'It is becoming increasingly clear that the present government firmly believes the Highlands are of no account and are simply not worth bothering about.'[38] Another leader in the *Inverness Courier* referred to the 'murder' of the Board[39]; Highland MPs called repeatedly in the House of Commons for the Board to be allowed to continue with its schemes; the Scottish Unionist MPs told the Secretary of State they opposed the merger;[40] the Highlands and Islands Advisory Panel stated its opposition;[41] and all the while the Secretary of State held his peace. He appeared to be dithering.

Evan Barron wrote him an open letter and printed it in full in the *Inverness Courier* on 21 May 1963. Reminding Mr Noble that he and his newspaper had always backed the Unionists, Barron credited himself with having started in 1928 the campaign that resulted in the birth of the Hydro Board, elaborated the grounds on which he felt able to speak for the Highlands ('I know ... the Highland people ... probably better than any other man living ... I have been fighting for the Highlands practically all my life'), reminded the Secretary he was Highland himself and ought to know the antagonism of the Lowlands, said the proposal would revive the old hatred between landlord and tenant, and went on in similar vein at some length. It was a magnificent piece of rhetoric and mounted in feeling with every line. 'For heaven's sake act now and put an end to ... these dangers by the simple act of announcing that there will be no merger of the two Boards ...' Barron told Noble that the retired Tom Johnston was 'nearly broken hearted' by the way the Mackenzie Committee had treated the Board and that the Secretary's silence was making the old man's health worse.[42] At last, in July 1963, Michael Noble announced in the House of Commons that the Hydro Board was safe.[43]

Three hydro schemes lay on his desk waiting for approval – for Glen Nevis, Fada-Fionn, and Loch a'Bhraoin. Meanwhile, the Strathfarrar scheme had been opened by Princess Alexandra on 7 June. The Board asked for inquiries into four new schemes, the three mentioned above and the Loch Laidon scheme, in October; and in November the Secretary of State initiated the inquiry into the Fada-Fionn and Loch Laidon schemes, scheduling it to start in the following January.[44] The Glen Nevis inquiry was put back on the shelf until the construction of the new pulp mill at Corpach should be completed in two years' time; the Glen Nevis scheme never saw the light of day again. John Alexander Dick QC was appointed to chair the inquiry, and Archibald Duncan Campbell, professor of applied economics at St Andrews University, was given the role of reporter.

Evan Barron smelled a rat. The Secretary of State, he said in an editorial when the inquiry began,[45] had instructed Dick and Campbell to take economic arguments into account, something never before done by a public inquiry into a hydro scheme and an aspect never intended to be covered by the 1943 Act. 'The Act was passed to benefit the Highlands and Islands, not the Treasury,' was Barron's comment. It seemed as if Michael Noble, scared by the outcry against the Mackenzie Committee's proposed merger, was nevertheless after trimming the Hydro Board's sails by more indirect means. The Secretary later attempted to correct any wrong impression by stating that, although the social obligations of the Board made it necessary to accept a lower return on its total investment than would be the case with a strictly economic enterprise, it was necessary by implication that the return on major new assets should be higher.[46]

The Board had published the details of the Fada-Fionn and the Loch a'Bhraoin schemes in March 1963. The intention behind the first was to harness the waters of two lochs, Fionn Loch and Lochan Fada, lying in a trough to the north of Loch Maree, to feed a power station near Furnace on the north shore of Loch Maree itself. To minimise the impact on the high scenic beauty of the area, the power station and aqueducts would be built underground and the transmission lines would be screened to hide them from the main road on the south side of Loch Maree. Loch a'Bhraoin, lying a short distance to the north of the Fada-Fionn complex, would be dammed and its water fed through a pipeline to a power station near the junction of the Cuileig river with the Droma in Strath More at the head of Loch Broom. The Fada-Fionn scheme was estimated to generate 51,000 kW and the Loch a'Bhraoin scheme 14,000 kW, and together the construction was expected to cost £8.8 million.

Twelve objectors, mostly owners of estates or fishery rights in the Wester Ross area, came forward to present evidence to the inquiry. In the face of strong arguments in favour of thermal generation, the Board fought to defend its plan to develop water power. At one point, an economist from Oxford University argued that the Board should never have been formed, as none of its schemes had ever been economically justified.[47] The inquiry lasted with breaks for almost three months – it ended on 24 March 1964, and it was another twenty months before the outcome was published. By that time there had been a General Election: the Labour Party had won, and Michael Noble had been replaced by Willie Ross as Secretary of State. Some other important events had also taken place. Michael Noble had paid tribute to the work of the Hydro Board at the opening of an international congress in Edinburgh of the International Commission on Large Dams, attended by a thousand engineers from fifty-three countries – the speech may have caused a few wry smiles in the Board headquarters.[48] The new Labour Government had announced its intention to set up the Highlands and Islands Development Board; and both Evan Barron and Tom Johnston had died.

Highland MPs repeatedly questioned the new Secretary of State in the House of Commons when they might expect to hear a decision on Fada-

Fionn, and repeatedly Willie Ross put them off with excuses and hints. In April 1965, Ross assured the House that the Hydro Board's future was not in doubt. In October the Queen opened the Ben Cruachan scheme, and in November the new Highlands and Islands Development Board held its first meeting. Finally, on 29 November, the Secretary announced that the Fada-Fionn and Laidon schemes were not to go ahead, because it was possible to generate electricity at less cost per kilowatt by other means. He left open the possibility that the schemes might be implemented in the future, but the social clause had not been strong enough on this occasion to save the day. The leader in the *Inverness Courier*, now with Evan Barron's niece, Eveline, at the helm, was thankful that Tom Johnston had not shared such a narrow actuarial view.[49] A few weeks later, Willie Ross opened the Board's new £11.5 million oil-fired power station, Carolina Port B, in Dundee and said there was still a place for hydro-electric power.

The Loch Awe and the Foyers schemes were the last of the major construction schemes undertaken by the Board. Loch Awe had three parts to it – the plan foresaw two conventional generation operations, at Nant and Inverawe, but also included a major innovation. The power station in the Ben Cruachan massif, dominating the northern shore of Loch Awe, was to be the first major reversible pumped storage operation. Reversible pumping had been installed at Sron Mor in the Shira scheme but there it was used to pump water from a lower (Lochan Sron Mor) to an upper reservoir (Lochan Shira) up a height of only some 130 feet; and the feature was only a small part of a much larger, conventional scheme. Ben Cruachan was to be on an altogether greater scale. The idea of pumped storage had been around for a long time – Sir Edward MacColl had considered it for Loch Sloy but its implementation had always been ruled out by the high cost. The central concept was to use off-peak electricity to pump water from the lower reservoir back up to the higher one when demand for electricity was low; when demand was high, the water released from the higher reservoir flowed down to generate more power. Early pumped storage schemes needed separate pumps and generators, but an advance in design allowed the incorporation of the pump and generator into one machine. The new design could be fitted into a smaller machine hall and the switch over from pumping to generating, by reversing the water flow through the machine, could now be accomplished in a matter of seconds instead of minutes and allow the engineers to respond quickly to surges in demand for electricity.

The construction of the Cruachan scheme involved nothing less than the

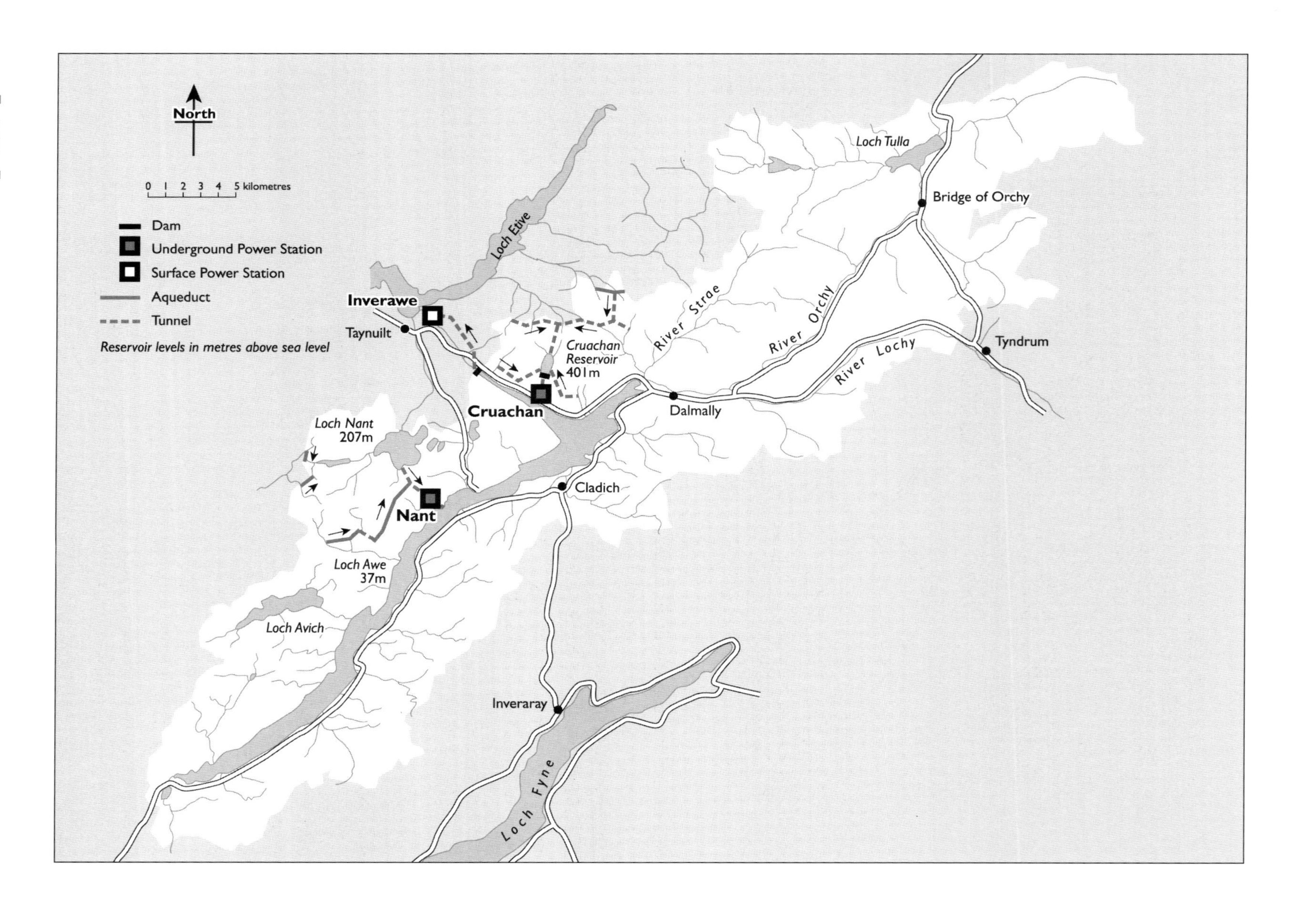

North
0 1 2 3 4 5 kilometres
Dam
Underground Power Station
Surface Power Station
Aqueduct
Tunnel
Reservoir levels in metres above sea level
Loch Etive
Inverawe
Taynuilt
Cruachan Reservoir 401m
Cruachan
River Strae
River Orchy
River Lochy
Loch Tulla
Bridge of Orchy
Tyndrum
Dalmally
Loch Nant 207m
Nant
Cladich
Loch Awe 37m
Loch Avich
Inveraray
Loch Fyne

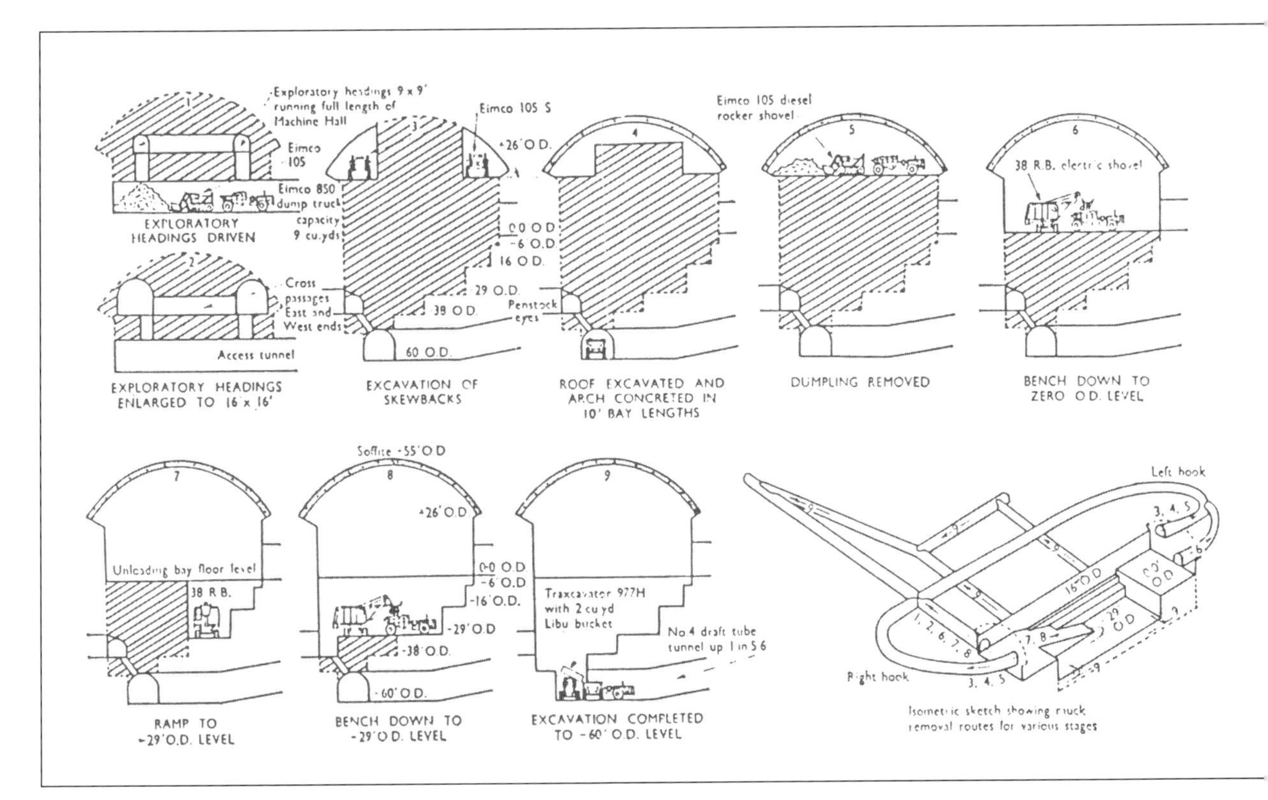

Fig. 5.

Above. The steps in the excavation of the machine hall inside Ben Cruachan (*reproduced with permission from the* Proceedings of the Institution of Civil Engineers).

Map 9.

Opposite. Awe

creation inside the mountain of a warren of tunnels and spaces for machinery, between the upper reservoir stored behind the dam to be built in the corrie on the summit plateau and the lower reservoir – Loch Awe itself. The Board awarded the contract as consulting engineers to James Williamson and Partners. George Rennie had been working for Williamson for five years on the Killin scheme in Breadalbane before he came to Cruachan in 1960 as resident engineer.

Work began with the construction of the access road from Loch Awe village up to the dam site. The camp sites and the accommodation for the staff were established and then work began to build an access tunnel from the shore of Loch Awe into the heart of Ben Cruachan. Trial bores had showed that the tunnellers would have to pass through mica schist before encountering granite. The aim was to site the machine hall with the generators deep in the heart of the granite but it took longer to reach this hard rock than had been expected and the machine hall was built closer to the edge of the granite than had been originally planned.

From the end of the access tunnel, two short shafts were driven vertically upwards to the level of the arch of the machine hall roof. 'Then we drove two horizontal tunnels, horizontal headings, along the line of the roof,' explained George Rennie. 'These were exploratory. We could see where the cracks were, the fissures and faults. They weren't too bad. The problem was

Plate 96.
The excavation of the machine hall in the heart of Ben Cruachan (*Edmund Nuttall Ltd*).

Plate 97.
The generating sets being installed in the machine hall in Ben Cruachan (*Edmund Nuttall Ltd*).

which way to go without spending a lot of time probing. In fact we only found one major fault and decided we could cope with it, although it turned out to be worse than it appeared to be at first sight.'

From the end of the access tunnel, the drillers excavated a major tunnel to the left that was to become the transformer hall, and then drove a further long tunnel, the 'left hook', curving up to roof level to meet the further ends of the exploratory headings. An equivalent but shorter tunnel was later driven to the right, the 'right hook', up to the level of the machine hall roof to allow more access for the removal of spoil – this tunnel was to be converted into the access to the visitor's gallery. Tunnelling inwards from the horizontal exploratory headings, the men carved out the space immediately under the roof of the machine hall. The arched vault of the roof was concreted and then the excavation of the machine hall began downwards in what was essentially a quarrying operation.

Tunnels were driven upward from the machine hall area vertically to form a cable and ventilating shaft, and at an angle of 55 degrees in the direction of the dam for the high-pressure shafts through which the water

Plate 98.
The access tunnel to the complex inside Ben Cruachan (*Edmund Nuttall Ltd*).

would flow to and from the upper reservoir. 'The shaft from the machine hall to the corrie is 55 degrees because at that angle, when building, gravity assists in getting rid of the spoil – the shaft is self-mucking,' said George Rennie. 'If it had been too steep you couldn't walk up it. We did in fact have ladders put in but you could make your way up the shaft without them.'

Barry McDermot, from Letterkenny in Donegal, was working in Paddy Gallagher's International Bar, a famous Clydebank pub, when he heard about the tunnelling at Cruachan, 'how good the money was and all that' and went north for a piece of the action. 'The first job I got was on the outfall,' he said. 'I was working with my brother Lawrence and my brother-in-law John Mulhern. It was long hard work, twelve-hour shifts, you started at seven in the morning until seven at night. You had a packed lunch given you in the canteen – cheese, corned beef and that crack. I was there, then I moved into the tunnel, with the same crowd, Edmund Nuttall.'

As Barry was small and light – he had earlier considered a career as a jockey – he was given the job of oiling the spaces in and around the moving shutter before concrete was pumped in to line the tailrace tunnel. The shutter

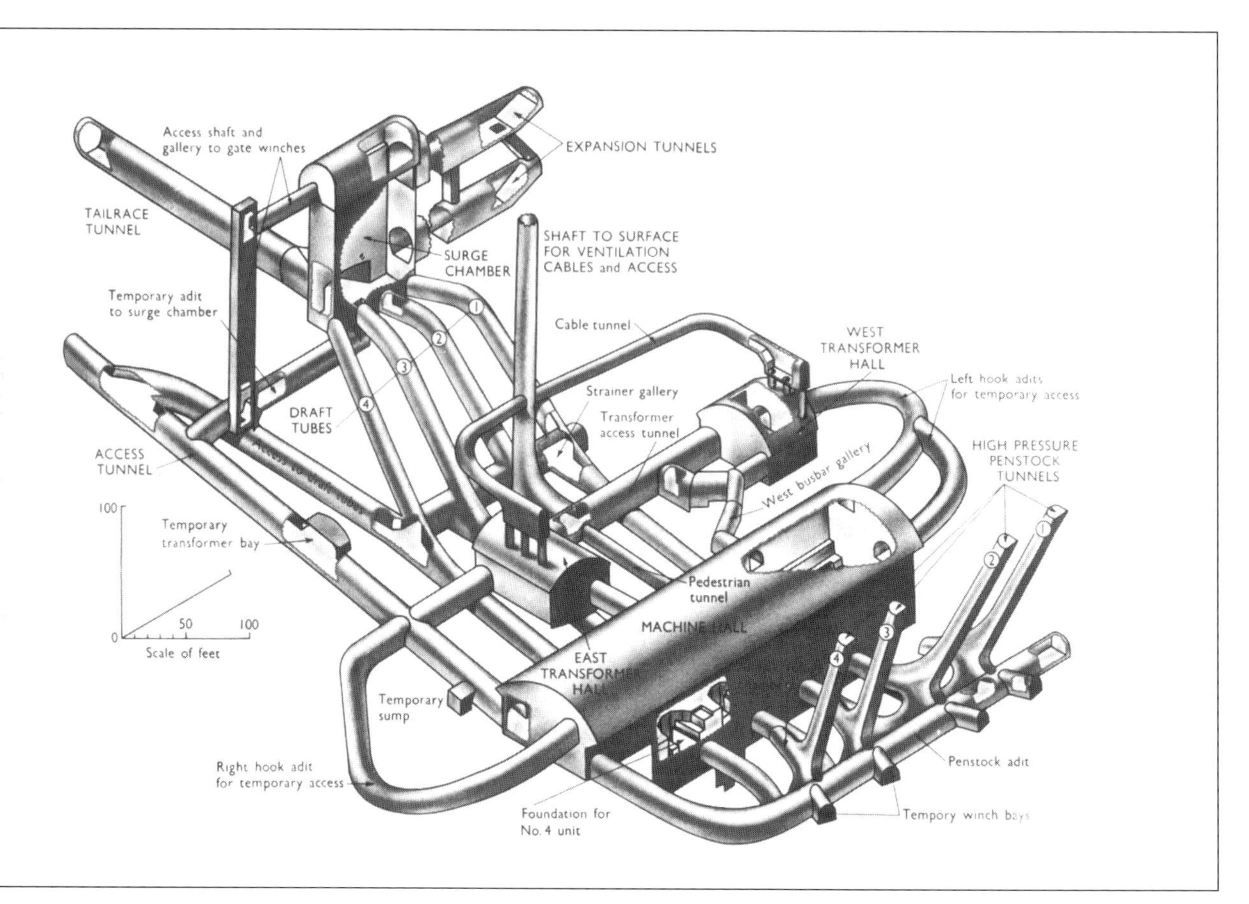

Plate 99.
This diagram shows the complex of tunnels and chambers constructed within Ben Cruachan. At the lower right, the four high-pressure penstock tunnels lead down from the dam and reservoir on the summit of the mountain to the generating sets in the machine hall. The tailrace tunnel and the access tunnel, on the left, lead out to the waters and shore of Loch Awe (*Edmund Nuttall Ltd*).

was pulled along rails by a winch. 'It was all rush, rush, rush. Then the shift boss says "I've got a job for you up in the [55 degrees] shaft". I says that'll be fine, so I went up to work in the shaft. I had to make sure there were no pieces of rock loose. That was a good job, I had that for about three months. I had to check the walls and chip away the loose stuff with a wee hammer, before it was concreted later.'

'Building the shafts was a bit of a headache,' said George Rennie. 'During the early stages we heard about a machine called an Alimak that had been developed for use in coal mines away up in the north of Sweden. We went up there to have a look at it and came to the conclusion it would be a very useful machine to have, or it could be made into a very useful machine.' Iain Macmaster remembers the team of Swedes who came over with it to instruct in its use. It was installed at Nant where the Swedes intrigued their hosts by sunbathing in lederhosen and drinking coffee laced with brandy from a two-gallon urn.

To use the Alimak, the tunnellers had first to reach the bottom of where the shaft was to be by conventional means and excavate a short distance upwards. Then two-metre lengths of racked rail were rock-bolted vertically

Plate 100.
Barry McDermot with his brother Lawrence and brother-in-law, John Mulhern, at Cruachan (*Barry McDermot*).

to the side walls. The original Alimak was a device about the size of a telephone kiosk that could hold two men. It had an air motor, and a wheel that engaged on the racked rail so that it could climb upwards to allow work on the rock above. When the rock was ready, the machine was brought down and withdrawn to one side to allow the roof to be blasted. Then the machine could be swung out again to ascend, and the men could take with them more lengths of rail to extend the trackway upward. 'We brought one over and it was used first in a short vertical shaft', said George Rennie, 'and it was highly successful. Ten times quicker than the conventional methods. Six to eight feet of drilling could be done each time, and the fallen rock could be cleared away from the bottom at the same time.'

The original Alimak worked in a shaft only four feet in diameter, excavating about one ton of rock at a time. The main shaft at Cruachan was to be five times bigger but the contractor, Edmund Nuttall, developed a bigger and better machine on the Alimak principle to permit excavation of the full-sized shaft.

Something like 330,000 cubic yards of rock were eventually to be removed from Cruachan's innards in the construction of the scheme. The machine hall became an immense, cavernous space with a floor area the size of a football pitch and a vertical height of 120 feet, enough for a seven-storey building. The buttressed dam is 1,100 feet long and 153 feet high.

The opening ceremony of the Ben Cruachan scheme took place on 15 October 1965. 'The date had been set about two years previously to accommodate Her Majesty the Queen', said George Rennie, 'but at the same time there was pressure to get the power station in service. Work was running late. I was in charge of getting the station into service and at the same time making provision for over 700 guests. Seating was arranged on scaffolding in the machine hall, and thirty-odd buses had to be laid on to convey the guests inside, all without interfering too much with the construction.' The *Oban Times* reported 'feverish activity' on the eve of the opening ceremony.[50]

The royal train arrived at Dalmally on the 15th, a blustery, showery morning, and the Queen was first driven up to see the dam and cross the spillway before joining the assembled guests in the machine hall. A faulty address system prevented most of the assembled throng from hearing all that was said during the speeches. Lord Strathclyde invited Her Majesty to start the first machine to inaugurate the scheme. 'Once she had pulled the switch to operate the set,' reported the *Oban Times*, 'the Queen turned round and saw reflected by many lights on a ceiling panel the machine building up its revs.'[51]

That wasn't quite the reality. 'We were supposed to run the machine but what we did was turn it with an electric motor, make all the noises of water running, valves opening, the hiss of compressed air, and so on – we faked all that,' said George Rennie. 'No water came down. We put a spotlight on the top of the machine that flashed on to the ceiling and, as the machine turned, the audience could see the light moving. To all intents and purposes that was the machine started up. There was great applause. Maybe a couple of dozen people knew what was really happening. About two months later the machines were properly ready.'

The Cruachan scheme was a great success. Its first major test in responding quickly to heavy demand for electricity came in 1966 at the time of the World Cup. It seemed as if the whole country was watching television when England played Germany in the final at Wembley on Saturday 30 July. The match went into extra time and the anticipated shut-off did not happen. Over a period of about an hour after the expected end of the game the demand for electricity yo-yoed and Cruachan played a crucial role in balancing the grid. As a result, more pumped storage schemes were planned. (Two were built in the Welsh mountains – at Ffestiniog and Dinorwic.) The

Board turned its thoughts back to Loch Sloy, a scheme Sir Edward MacColl had originally considered suitable for pumped storage, and engineers from Cruachan had already begun exploratory work when test bores revealed unsuitable rock and attention was diverted to a new site. The Board thought it better to go for a new scheme on the shoulder of Ben Lomond.

The engineers called it the Craigroyston scheme and envisaged an upper reservoir in a valley on the north-western slope of Ben Lomond and a power station on the eastern shore of Loch Lomond, nearly 1,000 feet below and almost opposite Loch Sloy. The scheme would be close to the industrial heartland of the Clyde. The planned capacity was 1,600 mW but in the long-term this could be increased, it was felt, to 3,200 mW. The Board publicised the plan with great attention to how the public might react to this tampering with Ben Lomond and emphasised that the rockfill dam would face north-east, and it and the reservoir would be all but out of sight from the other side of Loch Lomond; all the rest of the works would be underground. Construction was expected to take up to eight years, including two years for building access roads, and would employ a maximum workforce of 1,000.[52]

The Craigroyston scheme was never submitted for formal approval but in November 1969 construction of what was to be the Board's last large scheme began at Foyers.

Five weeks or so before the Queen opened the Ben Cruachan power station, Tom Johnston died – on Saturday 4 September 1965 at his home in Milngavie. He was eighty-three years old. The provost of Inverness said 'No words can say how grateful we are to him.'[53] The Highlands suffered another loss only a few months later. On 24 January 1966, the main north daily, the Aberdeen *Press and Journal*, carried the shocking news that 'Willie Logan is dead, and the whole of the Highlands are in mourning.' Two days before, shortly after half past ten on a misty Saturday morning, a chartered Piper Aztec carrying the contractor hit the pine-clad slopes of Dunain Hill on the western approach to Dalcross airport at Inverness. Ironically Willie Logan himself had been planning to try for his pilot's licence but on this occasion the plane was being flown by ex-Squadron Leader Peter Tunstall, who suffered injuries but survived the crash. Willie, his sole passenger, was killed. Many in the north can still remember what they were doing when they heard the news. Bill Mackenzie was shopping in Inverness and had just gone through the door of Woolworths when somebody asked him if he knew what had happened. Laurie Donald was in Glasgow at the time, working on a Logan contract, and on the Saturday afternoon the managing director of the

Plate 101.
A diagram of the Foyers scheme superimposed on a photograph of the landscape (*Edmund Nuttall Ltd*).

Glasgow office telephoned with the news. The pews in the Free Church in Dingwall, reputed to seat 1,000, were packed for Willie Logan's funeral and probably a thousand more people listened to the service outside on loudspeakers.[54]

Duncan Logan Construction Ltd, Willie's firm, had won the contract for the so-called upper part of the Foyers scheme. In 1967 the Hydro-Electric Board had taken over the British Aluminium plant at Foyers when the aluminium smelter closed. K. R. Vernon, then the Board's general manager and chief engineer, recognised the site's potential for pumped storage and in the following year the details for the Foyers scheme were made public. None of the old BA structures would be used: a new pipeline and tunnel system would be built to lead water down a head of 589 feet from Loch Mhor to the east

Plate 102. Excavating the surge chamber on the Foyers scheme (*Edmund Nuttall Ltd*).

of Loch Ness to a power station on the shore of the famous loch itself, which would act as the lower reservoir. The catchment area of Loch Mhor would be enlarged to eighty square miles by drawing water from the River Fechlin and the River E, both of which eventually drained into Loch Ness in any case. Initially the cost was put at £10.6 million. The output was expected to be 307 million units from pumped storage and 93 million units from conventional flow. The construction began in the autumn of 1969.

The scheme began with an upper works, made up of a low-pressure tunnel and an intake from Loch Mhor. The tunnel was 22.75 feet in diameter and ran for 9,000 feet to the top of the high-pressure shaft that marked the beginning of the lower works. The high-pressure tunnel dropped vertically for 370 feet to a horizontal tunnel that led to the power station. In the power station two vertical shafts, each 165 feet deep and 62.5 feet in diameter, housed the two 150 mW generating sets. A leaflet published by the Board

Plate 103.
Constructing a tunnel on the Foyers scheme (*Edmund Nuttall Ltd*).

Map 10.
Opposite. Foyers

showed that the Scott Monument on Princes Street could easily fit into each of these shafts, and added the interesting information that the turbines would in fact be sited 114 feet below the level of the surface of Loch Ness. At full load the turbines would pass 152 million gallons of water per hour into the loch, and when pumping would draw from the loch 123 million gallons per hour. The whole system would be remotely controlled from the existing Garry-Moriston control centre at Fort Augustus.

Laurie Donald returned to work on the upper works for Willie Logan: 'It was on a massive scale with a huge intake works on Loch Ness side. No sooner had we started than the Logan firm went bust – that was in 1970. Edmund Nuttall had the contract for the lower works – they had originally tendered for the upper works as well and now they went back to the Board and said we can carry on as if we had won both tenders. The Board agreed. I became an employee of Nuttall and ended up running the lower works. That post used to be called agent, but by this time we were being called project managers.' The liquidation of the Logan company was a sad blow to the Highland economy but not a surprise to those observers who felt that without the driving presence and forceful personality of Willie it had become dangerously overstretched. Nuttall took over the whole scheme in May 1970.

A number of technical problems had to be overcome in the construction of the Foyers scheme. The rock between Loch Mhor and the surge shaft was highly fractured and this, combined with the need to cross Glen Liath, meant part of the distance had to be covered by a surface pipeline. The 275-foot surge shaft rising from the end of the low-pressure tunnel had a diameter of

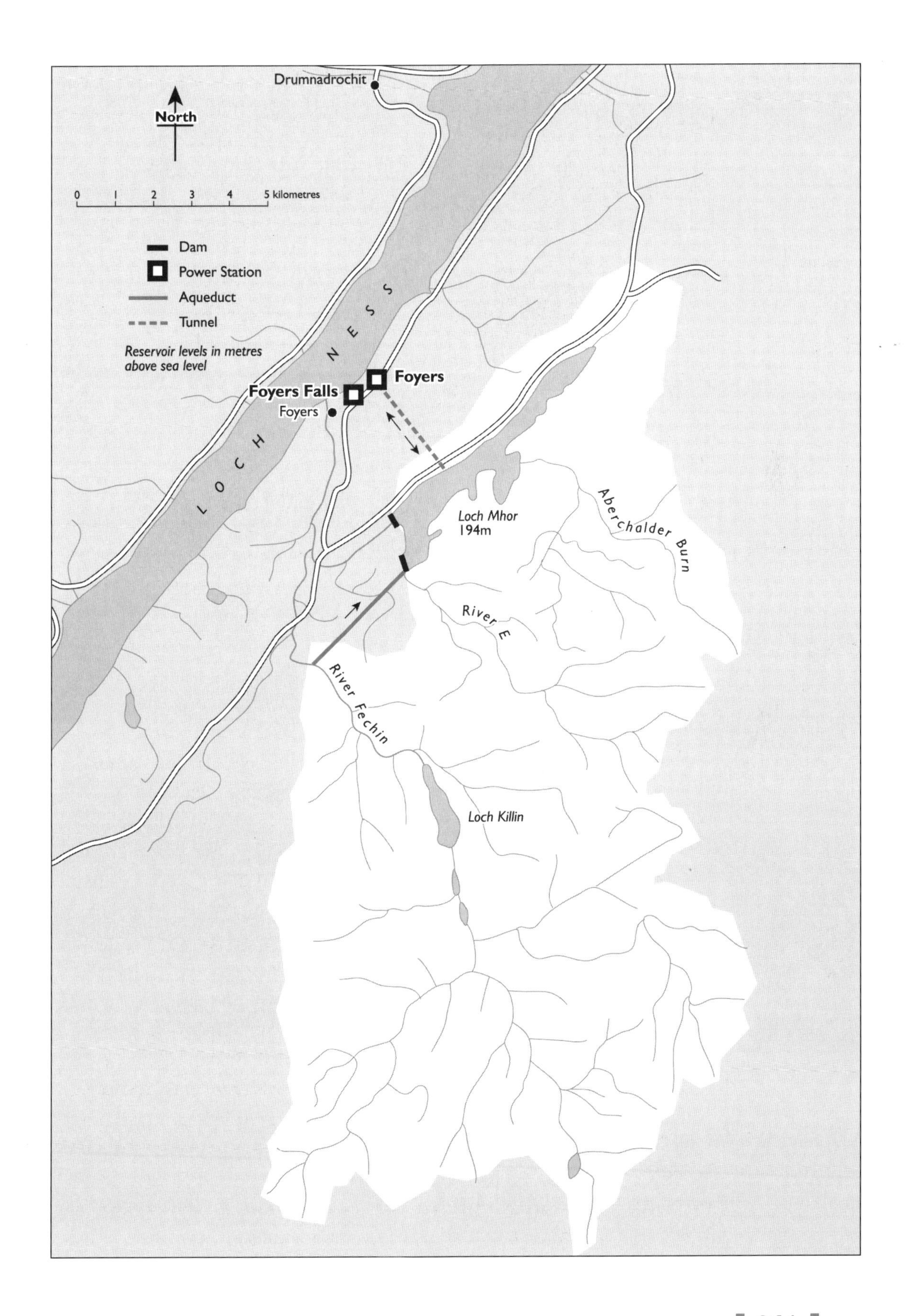
Drumnadrochit
North
0
1
2
3
4
5 kilometres
Dam
Power Station
Aqueduct
Tunnel
Reservoir levels in metres above sea level
LOCH NESS
Foyers
Foyers Falls
Foyers
Loch Mhor
194m
Abercholder Burn
River E
River Fechin
Loch Killin

sixty-one feet and a hydraulic, continuously climbing shutter was installed to allow the placing of the concrete lining – the whole operation took only twenty-one days. Despite the problems from the unhelpful geology and the need to re-schedule various stages of the work programme, the scheme was completed more or less on time. The final cost, however, was £20.2 million, nearly twice the estimate – although most of the increase was due to inflation.

Willie Ross, Secretary of State, officially opened the Foyers scheme on 3 April 1975. About 100 guests assembled in the two great shafts beside the generators and the Royal British Legion Pipeband from Inverness were on hand to enliven the occasion. Times had changed, however, and there was an undertow of sadness about the whole event, as if everyone sensed they were witnessing the last act in a great enterprise. Oil had been discovered in the North Sea, promising wealth and energy. 'There is no significant potential left for conventional hydro stations in Scotland,' said the Secretary of State 'Even if the conservationists would go along with such minor opportunities as still exist, the whole scale of electricity generation has so expanded that these would be a mere drop in the bucket. The scope for further hydro generation is undoubtedly in the field of pumped storage, which this Board has the experience and resources to develop.'[55] But that was not to be.

One major part of Tom Johnston's dream for the Highlands did not materialise to the extent he hoped for, although it was not for the want of trying on the Board's part. In the end, industries sprang up only to a limited extent, as geographical realities proved stronger than the availability of electric power. With the visions of small enterprises came the idea of 'model' villages, each with some 2,000 inhabitants, in the glens. The provost of Elgin described in 1947 how such a village could be founded in Glen Affric but other members of the Local Authorities Committee scoffed at the notion.[56] Depopulation was still continuing as the first hydro schemes were springing from the river beds – the Highlands and Islands lost almost 4,000 emigrants in the two years after the end of the Second World War.[57] In 1950, the Board announced its intention to create a model village at Cannich – it had similar plans for Contin, Fort Augustus and Killin – where the new energy would support farms, tourist ventures and small businesses.[58] The Board built houses in several villages but, on the whole, the thriving businesses did not appear.

From time to time it seemed as if the Board's efforts to attract industry were having some success, for example when Ferranti Ltd opened a transformer factory in Inverness and the General Electric Company were indicating an intention to set up an electric cooker assembly plant,[59] but overall not a great deal was achieved. The population of most of the seven crofting counties, as the bulk of the Highlands and Islands was sometimes labelled, continued to fall throughout the 1950s: the 1961 census revealed a drop of 4.6 per cent, just under 12,500 people, the only area to escape this drain being Caithness where the siting of the Dounreay nuclear power station had brought about a massive influx of workers and their families and had boosted the county's population by 20 per cent.[60]

Lord Strathclyde assured a press conference in Edinburgh at the beginning of 1966 that there was no question of the Board slacking in its efforts to improve the economic prospects of the Highlands.[61] He was forced to admit later that year that the Board had failed to stop the drift of people from the region but argued that they had achieved results 'far beyond the dreams of the most enthusiastic supporter' of the original Bill. In Birmingham in November 1967, for example, the Board mounted a campaign to attract industry but most firms did little more than express polite interest. The chief commercial officer of the Board stated, however, that since 1943 250 industries had started in the north and were now employing 16,000 people.[62] In 1973 the Board conducted a dozen English businessmen on a two-day visit to the Highlands, the latest in a series of such tours.[63] The Board's 'Key to Freedom' campaign had its first success later that year when a subsidiary of the Morgan Crucible Company announced its plan to transfer to a new factory in Perth.[64]

The Highlands and Islands Development Board claimed more success: its efforts to create employment had checked depopulation, it announced, in June 1968. The 1971 census confirmed this, recording a population increase of 6,000 in the previous decade; good news, but the phrase 'Highland problem' was still in use. Professor Robert Grieve, the HIDB's first chairman, used the term in the organisation's first report[65] and incidentally also acknowledged the pioneering inspiration of the Tennessee Valley Authority. The HIDB also recognised the efforts of the Hydro-Electricity Board and the latter's chief commercial officer, J. C. N. Baillie, joined the HIDB Board in its early years.

The Board could point to its success in tourism, an industry already worth £80 million a year by 1961. When the Board opened the doors for six days at some power stations to mark its twenty-fifth birthday, over 25,000 visitors called by. The Cruachan station welcomed 9,000 visitors in the summer of

1969, and around 10,000 were paying a call on the Pitlochry power station and fish ladder every year.[66] Pitlochry now has a small exhibition describing how electricity is generated and visitors can also see the interior of the fish ladder, and salmon and trout passing through.

In 1972 Cruachan was equipped with a visitor's centre and it has remained a popular attraction since. Scottish Power took over the power station in 1991 and regard it proudly as a showpiece. In summer, coaches call and tours depart every half hour to bring visitors into the heart of the mountain. Around 50,000 make the trip every year, and learn from the guides how the whole operation works. An electric bus brings them down the access tunnel, two-thirds of a mile through dim rock walls to a viewing chamber high above the generating hall. This far inside, the temperature is steadily mild and subtropical plants flourish in pots under special lights. The rock is dark, almost black, with white splotches where seeping water has left mineral deposits and matchstick-sized stalactities sprout from the roof. The generating hall is vast: the length of a football pitch and 127 feet high. Four gleaming turbines sit in a line, and the whole area is tiled and clean. A large wooden mural by Elizabeth Faulkner illustrates the story of Cruachan, a witch who is reputed to have guarded a spring on the summit of the mountain. In its centre, a Celtic cross bears the names of the fifteen men who lost their lives in the construction.

The Board attached great importance to the appearance of the schemes. Little could be done to disguise the massive intrusion of dams but the power stations and some other structures made use of local stone. This revitalised the moribund quarrying industry and the declining need for stone masons in some parts of the Highlands. James Shearer, one of the Board's panel of agricultural advisers, suggested to Edward MacColl that local stone and slate would improve the appearance of the Nostie Bridge power station and thereafter it became standard policy to use traditional materials wherever possible.[67] Yellow sandstone from Burghead and red sandstone from the Tarradale quarries on the Black Isle were used to build in the Beauly-Affric area. The Finlarig power station was built in Aberfoyle stone, and Mossford in Tain sandstone. This approach also made economic sense: building the diesel generating plant in Kirkwall from local stone saved the considerable sum that would have been spent on importing concrete and steel.[68] Occasionally the nod in the direction of tradition went further: the sculptor Hew Lorimer was commissioned to carve four Celtic beasts to grace the wall of Fasnakyle.

Over half a century on from their construction, almost all the scars of the dams' birth have been obliterated from the landscape and the dams

Plate 104.
Constructing the Orrin power station. Masons used stone to build the outside walls around the girder framework, November 1958 (*NOSHEB*).

Plate 105.
One of the Celtic animals carved by Hew Lorimer on the wall of the power station at Fasnakyle (*author*).

themselves have acquired a weathered look. The nine-mile road from Cannich to Mullardoch is still single-track, rising, falling, twisting to find a feasible gradient, and it still feels as remote as it did in the 1940s to the men who built the scheme. The immense concrete face of the dam still shows the shuttering lines and the marks of dumb bolts but it has also been colonised by moss and lichen, and the white concrete has become dark and discoloured. The Loch Sloy dam, the first one to be built for the Board, is dark with age, almost the same colour as the outcrops of bedrock beside it and, in mist, if it were not for the barn door-like arches of the buttresses, it might be mistaken at a distance for an unusual geological formation. Further down, the glen is webbed with pylons and power lines, a much more intrusive presence. Driving around the Perthshire hills, one can come suddenly on a power station nestling half-hidden in trees, the only man-made sounds the hum of the generators and the slow swirl of water in the discharge pool.

Opinions remain divided over the environmental impact of the large schemes. Sandy Payne now looks on Monar dam, where he worked as a chainman, with mixed feelings: there is pride in having helped to build it, joy in remembering seeing eagles, greenshank and the wildness of the hills for the first time, and sadness because the dam really changed the ecology.

To older hands, a hydro-electric scheme represents the generation of power, the creation of jobs, the opening of inaccessible areas, a modernising force that breaks a landlord's grip; but to a younger generation, with a more developed environmental consciousness, they stand for the destruction of nature, the intrusion of 'big' business into rural areas, the erosion of indigenous cultures. There is now greater attention being paid to renewable sources of energy, as the burning of fossil fuels threatens to bring on global warming with many unforeseeable consequences, and there is much talk of harnessing wind, wave and solar power. Somehow, hydro power is overlooked in these discussions although it is clean, renewable, and releases no toxic by-products and no waste. Ironically, the Hydro Board did experiment briefly with wind generation and a peat-fired power station in the 1950s but these proved uneconomic.

In 1989 the privatisation policies of the Conservative Government of Margaret Thatcher finally caught up with the Board. The North of Scotland Board and its South of Scotland neighbour were to be turned into two companies – Scottish Hydro-Electric plc and ScottishPower plc. At the time the North Board had 3,500 employees and an annual turnover of £348 million.[69] During 1989-1990, SHE took over most of the property, rights and liabilities of NSHEB and became a business empowered to sell electricity to

Plate 106 and 107.
Digging the trench for the underground pipeline on the Cuileig River scheme, summer 2001 (*author*).

Plate 108.
The underground pipeline on the Cuileig River scheme, summer 2001 (*author*).

industry and regional electricity companies anywhere in the UK. Its turnover in its first year was £435.7 million, and its profit £108.7 million. In the midst of this commercial makeover, many feared the social clause – the heart of Tom Johnston's vision – would be abandoned, if not simply overlooked. SHE stated it would make 'determined efforts to ensure that [the social benefits] shall not be [lost]'[70] and pointed to flood alleviation, arts sponsorship and conservation projects as examples of its good works. In their annual report for 1996, Scottish Hydro-Electric re-affirmed its parent Board's ethos: '[Our] roots are in the north of Scotland and serving our community there is our first priority.' By 1990, however, hydro power had become only a small part of the range of the new company's activities – only 17 per cent of its electricity output came from hydro generation, against 31 per cent from oil and gas, 23 per cent from nuclear power stations, and 27 per cent from coal-fired stations. Only 39 per cent of its revenue came from domestic consumers in the north of Scotland. In Peterhead it was modifying a 1,320 mW power station to burn North Sea gas and a new subsea cable was being planned to bring power from cheaper generation plants on the mainland to the Western Isles.

Work on the first hydro-electric scheme to be built by Scottish Hydro-Electric began in December 2000 on the Cuileig river in Wester Ross. This is the Loch a'Bhraoin scheme revisited, a slight modification of the one included as number seventy-seven in the Hydro Board's original list of 102 projects. The aim was to tap the water flowing from Loch a'Bhraoin down the Cuileig to join the River Broom in Strathmore, and the work was done in 2001. A weir on the Cuileig diverts water into a pipeline falling to a power station on the floor of the strath. The scheme cost £4 million and the main contractor was Miller Civil Engineering Services Ltd, with the Strathpeffer firm of Kenneth Stewart Ltd subcontracting for the pipeline.

There were many contrasts between this small scheme and the major efforts of the 1950s. The work camp was now a cluster of interconnected portable cabins. Only some forty men were involved in the construction and the turbine in the power station is remotely controlled. The pipeline is constructed from six-metre lengths of glass-reinforced plastic that were slotted together through rubber sleeves. Diggers grubbed out the trench and, depending on the amount of rock to be excavated, the pipeline was layed at a rate of up to fifty metres per day. Most of the shuttering on the weir was prefabricated and the old shuttering techniques were used only on odd corners. The environmental specification was rigorous: top soil was stripped and preserved for relaying, the pipeline is buried along almost all its length, access roads are sited so as not to impinge on the view from the Dundonell road, the amount of compensation water is clearly laid down, and the roof of the power station is covered in soil and planted with grass. Some things were the same as always: the drenching rain, the midgies, the mud. The engineers, explained that construction projects now are cut to the bone and there is no room for frills in the highly competitive tendering; the old Board's attention to design and the use of local stone has gone.

There have been in recent years a number of hydro-electric-generation initiatives in the Highlands. Some of these have been confronted by objections and have not proceeded, the casualty list including the Hydro Board's own run-of-river schemes for the Grudie and Talladale Rivers in 1983. The Assynt Crofters Trust, after a long period arguing with environmentalists, opened their own scheme in 2000; this joint project between the Trust and the Highland Light and Power company of Dundee cost £500,000 and taps the discharge from Loch Poll on the Stoer peninsula.[71] In July 2001, Brian Wilson, the energy minister and no stranger to the Highlands – he was one of the founders of the radical *West Highland Free Press* – announced a £250-million package to be spent over the next decade on refurbishing the hydro-electric power stations built by the Board. 'This is a major signal that hydro-electric power still has a huge part to play in the Government's strategy for renewable energy,' he said. 'The expansion of hydro was one of the great visionary acts of the postwar period.'[72]

The men and women who were intimately involved in the construction of the dams feel pride in what they did and sometimes feel that the extent of the Board's achievement is not fully appreciated now. Some see it as one of the great achievements of postwar Europe, fully comparable to other major construction programmes. Thirty-four of the Board's dams were on a scale to be included in the World Register of Large Dams, and the hydro schemes were frequently included on the programme for a state visit. Experts with a professional engineering interest came to see them, including from the Soviet Union and the United States – and, among them, came the director of the Tennessee Valley Authority, the body Tom Johnston sought in some ways to emulate. Lord Wilson, the chairman of Scottish Hydro-Electric, said on a BBC Radio Scotland programme to mark the fiftieth anniversary of the original Board in 1995, that he couldn't help feeling enormous admiration for the engineers and the workers who built the power stations in a time of great economic difficulty, adding 'The power lines across the Lecht were put in in snowstorms. It was a tremendous effort against great odds. We inherit their work and we must never forget what they did.' When I asked Patrick McBride in Donegal if he felt he had been doing something important when he was wrestling to control the Tummel, he replied in one word: 'Definitely.'

Notes and References

The following abbreviations have been used for the most frequently quoted Highland newspapers:

IC *Inverness Courier*
OT *Oban Times*
PA *Perthshire Advertiser*

Introduction

1. The figures are from MacColl, 1946, and the Cooper Report.
2. Don Smith.

1

1. Highlands and Islands Development Board Annual Report, 1983.
2. *The Highlands of Scotland*, 1936. Quigley, who was born in Stirling in 1895, was well known as a linguist, economist and writer. From 1931 to 1943 he served as chief statistical officer for the Central Electricity Board.
3. *Off in a Boat*, 1938.
4. Turnock, 1970.
5. Greenock and Godalming are the most frequently quoted places in this context, but a letter from Louis Stott, Dumbarton, to *The Scots Magazine*, March 1979, states that the first hydro-electric plant in Britain was installed at Cragside, Northumberland, in 1879-82, and the first commercial scheme was at Portrush, Northern Ireland, in 1883.
6. *North Star*, 26.11.1903.
7. Payne, the source of most of the information about pre-1945 hydro-electric developments; also the *Third Statistical Account, Perth & Kinross*, 1979
8. MacGill.
9. *IC*, 16.8.21.
10. *IC*, 10.1.30
11. *Northern Times*, 23.3.33.
12. *IC*, 11.2.38.
13. *IC*, 8.4.38.
14. In later times, Tom Johnston discreetly bought up all the secondhand copies of his book that he could find, and enlisted the help of bookdealers in this task.
15. Johnston, *Memories*.
16. Before Johnston went to London for his interview with the Prime Minister, Lady Churchill had visited Mrs Johnston in Glasgow. It is possible that some preliminary soundings were made on that occasion on Johnston's possible reaction to being offered the Secretary of State post.
17. *IC*, 12.9.41.
18. According to the *Dictionary of National Biography*, Lord Cooper remained a shy man in private life but he adored children and loved cats.
19. *IC*, 27.4.65.
20. *IC*, 27.4.65.
21. *IC*, 7.9.65. In the 1940s, almost the entire press corps were positively minded towards the Hydro-Electric Board, the major exception being the papers in the D. C. Thomson stable who were opposed to nationalisation. Even the Thomson papers came 'on side' at the time of the Mackenzie Report.
22. *IC*, 18.12.42.
23. *IC*, 23.1.43.
24. *IC*, 7.5.43.
25. Johnston, *Memories*.
26. Payne.
27. *IC*, 27.8.43.
28. *IC*, 10.9.43.
29. *IC*, 14.9.43.
30. The full list of 102 schemes is given in MacColl, 1946.
31. *IC*, 2.6.44.
32. *IC*, 31.3.44.
33. *IC*, 29.8.44.
34. *PA*, 19.7.44.
35. *PA*, 2.8.44.
36. *PA*, 16.8.44.
37. *PA*, 26.8.44.
38. *PA*, 13.9.44.
39. *PA*, 21.10.44.
40. *IC*, 17.11.44.
41. *PA*, 25.10.44.
42. *IC*, 14.11.44.
43. *IC*, 20.2.45.
44. *IC*, 18.12.45.
45. *IC*, 20.4.45.
46. *IC*, 20.2.45.
47. *IC*, 18.12.45.
48. *IC*, 15.2.46.
49. A full account of the engineering of the Loch Sloy scheme is given in Stevenson, 1952.
50. Quoted in Payne.
51. *PA*, 14.3.45.
52. *IC*, 27.4.45.
53. *PA*, 2.5.45.
54. *IC*, 27.4.45.
55. Quoted in Payne.
56. *IC*, 28.8.45.
57. *PA*, 24.4.46.
58. *PA*, 27.4.46.

59. Robert Hay, 'The shape of things at Pitlochry', *The Scots Magazine*, August 1949.
60. *PA*, 26.4.47.
61. *PA*, 5.2.47.
62. *IC*, 24.9.46.
63. A full account of nationalisation is in Payne
64. This account is based on information from Mackinven.
65. *IC*, 14.1.47.
66. *PA*, 19.2.47.
67. *IC*, 9.5.47.
68. *IC*, 3.6.47.

2

1. *IC*, 27.12.49.
2. *IC*, 24.11.53.
3. *IC*, 4.2.55.
4. *IC*, 19.7.57.
5. *IC*, 25.12.59.
6. *Third Statistical Account, Ross and Cromarty*.
7. *IC*, 4.2.55.
8. Information from Sandy Macpherson.
9. *IC*, 10.6.49.
10. *IC*, 27. 12. 49.
11. *PA*, 8.10.47.
12. The blondin was named after Blondin, the French tightrope walker who, in 1859, walked across the Niagara Falls.
13. *IC*, 18.5.48.
14. *IC*, 29.8.52.
15. *IC*, 7.4.59.
16. *Highland News*, 26.1.68.
17. *IC*, 7.4.59.
18. *IC*, 7.4.59.
19. *IC*, 15.5.64.
20. I am grateful to the Muir of Ord firm D. J. MacLennan and Son, who have kept a scrapbook of cuttings about the Logan company with details of Willie Logan's rise to prominence.
21. R. J. McLeod Ltd, with its headquarters in Glasgow, is still a thriving construction company.
22. *IC*, 24.12.48.
23. *IC*, 10.6.49.
24. *IC*, 6.3.51, 16.3.51.
25. *IC*, 24.7.51.
26. *IC*, 14.10.52.
27. Commander Ian 'Tich' Fraser won the Victoria Cross in 1945 when he commanded a midget submarine attack on Singapore harbour and sank the Japanese ship *Takao*.
28. *IC*, 6.2.53.
29. Concrete chemistry is a specialist subject. Portland cement, the most common type used in Britain, is made by heating a mixture of 75 per cent limestone and 25 per cent clay to 1,500°C to produce a clinker which is then ground with some gypsum to form the familiar grey powder.
30. *IC*, 2.10.56.
31. The Westminster estates in Sutherland extended to around 113,000 acres, according to John McEwan in *Who Owns Scotland* (second edition, 1981) and at the time of the Shin scheme they were much larger.
32. *Northern Times*, 9.7.54.

3

1. *PA*, 28.12.55.
2. *IC*, 5.12.50.
3. *PA*, 25.5.57.
4. *PA*, 1.6.57.
5. *IC*, 13.11.53.
6. *PA*, 31.8.49.
7. *IC*, 7.3.50.
8. *IC*, 2.11.54.

4

1. *IC*, 12.11.48.
2. Serious crime seems to have been rare. I heard the story of a drive-by shooting at one camp. A gambler called Docherty was found dead at the time of the Cuachan scheme but details were impossible to confirm.
3. *PA*, 24.11.48.
4. *IC*, 17.11.53.
5. La Scala projected its last reel in January 2001, but Gellions is still on the left side of Bridge Street, a few yards up the street from the Ness Bridge.
6. *IC*, 15.10.46.
7. *IC*, 1.3.49.
8. *IC*, 10.3.53.
9. *IC*, 23.1.53.
10. *IC*, 23.3.48.
11. *IC*, 6.12.49.
12. *IC*, 3.2.50.
13. *OT*, 27.1.62.
14. *IC*, 30.1.48.
15. *PA*, 10.11.51.
16. *IC*, 19.3.57.
17. *PA*, 3.12.55.
18. *IC*, 29.4.49.
19. *IC*, 16.8.49.
20. *IC*, 14.4.50.
21. *IC*, 23.11.51.
22. *PA*, 26.3.60.
23. *IC*, 28.4.53.
24. *IC*, 18.1.55.

5

1. Brochure published by the Hydro-Electric Board to mark the official opening of the Loch Sloy scheme.
2. *IC*, 20.10.50.

3. The source is Norrie Fraser (ed), *Sir Edward MacColl: A Maker of Modern Scotland*, Stanley Press, 1956, quoted in Payne.
4. *IC*, 15.4.52.
5. *IC*, 15.4.52.
6. *IC*, 7.9.54.
7. The Australian climate killed the trees.
8. *IC*, 22.7.49.
9. *IC*, 2.2.51.
10. *IC*, 4.5.51.
11. *IC*, 6.3.53.
12. *IC*, 20.6.47.
13. *IC*, 3.3.61.
14. *IC*, 11.4.52.
15. *IC*, 6.10.64.
16. *OT*, 29.4.65. A Belling Compact electric cooker cost £39 10s in 1965, and the better off might opt to fork out 75 guineas for a Moffat E75.
17. *IC*, 8.1.52.
18. *Third Statistical Account, Argyll.*
19. *Third Statistical Account, Ross and Cromarty.*
20. *IC*, 29.6.51.
21. *IC*, 7.10.52.
22. *Northern Times*, 20.11.53.
23. *Northern Times*, 29.1.54.
24. *IC*, 22.1.54.
25. *IC*, 9.12.52.
26. *IC*, 11.5.54.
27. *John O'Groat Journal*, 11.8.61.
28. *John O'Groat Journal*, 16.12.60.
29. *IC*, 17.2.48.
30. *IC*, 26.2.54.
31. *Northern Times*, 29.1.54.
32. *IC*, 22.2.55.
33. Hydro-Electric Board, Annual Report, 1954.
34. *IC*, 27.2.59.
35. *IC*, 3.3.61.
36. *IC*, 18.6.65.
37. *OT*, 21.4.66, 16.3.67.
38. *OT*, 7.7.66. A few far-flung corners remained without a mains supply for years: the sixteen residents of Glen Etive, where the one-pupil school had gas lamps, were finally due to be connected in 1981 (*Glasgow Herald*, 19.5.80).
39. *IC*, 1.8.68.
40. *IC*, 13.6.69.
41. *IC*, 29.6.71.
42. *IC*, 23.5.72.
43. The experimental fast breeder reactor built at Dounreay in Caithness was intended to supply cheap electricity.
44. The dangers associated with the erection and installation of power lines led occasionally to fatal accidents, mainly through falls or blows from equipment.
45. *IC*, 19.4.55.
46. Payne.
47. The quoted prophecies are given in Elizabeth Sutherland, *Ravens and Black Rain*, London, 1985.
48. *IC*, 3.6.47.
49. Cameron McNeish, *The Wilderness World of Cameron McNeish*, Edinburgh, 2001.
50. *IC*, 18.6.48.
51. Seton Gordon.
52. J. Morton Boyd, *Fraser Darling's Islands*, Edinburgh, 1986.
53. Murray.
54. Simpson.
55. *IC*, 24.12.71.
56. *IC*, 9.5.72.
57. *IC*, 31.10.72.
58. *IC*, 18.7.52.
59. Iain R. Thomson, *Isolation Shepherd*, Edinburgh, 2001.

6

1. See, e.g., D. Mills and N. Graesser, *The Salmon Rivers of Scotland*, London 1981.
2. Payne.
3. *IC*, 22.5.51.
4. *Highland News*, 9.6.66.
5. *IC*, 2.10.56.
6. *IC*, 17.11.59.
7. *IC*, 13.12.57.
8. *IC*, 6.6.58.
9. *IC*, 30.6.50.
10. *IC*, 12.7.55.
11. See, e.g., *IC*, 29.2.52.
12. *IC*, 18.7.58.
13. *IC*, 17.2.56.
14. *IC*, 26.10.56.
15. *IC*, 19.7.57.
16. *IC*, 6.6.58.
17. *Highland News*, 30.4.70.
18. *IC*, 5.10.71.
19. *IC*, 28.11.58.
20. *IC*, 27.2.59.
21. *ibid.*
22. *Highland News*, 3.3.61.
23. *IC*, 13.6.69.
24. *Highland News*, 23.2.62.
25. *IC*, 8.6.67.
26. *ibid.*
27. *IC*, 29.12.59.
28. *John O'Groat Journal*, 11.11.60.
29. *IC*, 13.5.60.
30. Payne.
31. *PA*, 28.12.60.
32. *John O'Groat Journal*, 17.3.61.
33. *IC*, 8.12.61.
34. *John O'Groat Journal* 11.8.61.
35. *Highland News*, 9.11.62.
36. *IC*, 25.1.63.
37. *IC*, 2.4.63.
38. *ibid.*

39. *IC*, 19.2.63.
40. *IC*, 29.3.63.
41. *IC*, 23.4.63.
42. *IC*, 21.5.63.
43. *IC*, 12.7.63.
44. *IC*, 5.11.63.
45. *IC*, 7.1.64.
46. *IC*, 10.1.64.
47. *IC*, 25.2.64.
48. *IC*, 5.5.64.
49. *IC*, 30.11.65.
50. *OT*, 14.10.65.
51. *OT*, 21.10.65.
52. Hydro-Electric Board booklet, 1977.
53. *IC*, 7.9.65.
54. There was a rumour that Willie Logan had been piloting the aircraft himself on the fatal flight. Tunstall's testimony rules this out but it was quite in keeping with Willie's character that he might have taken the controls.
55. *IC*, 4.4.75.
56. *IC*, 27.6.47.
57. *IC*, 9.3.48.
58. *IC*, 29.9.50.
59. *IC*, 3.9.57.
60. *IC*, 24.5.63.
61. *IC*, 4.2.66.
62. *Highland News*, 7.12.67.
63. *IC*, 16.3.73.
64. *IC*, 24.7.73.
65. Highlands and Islands Development Board, Annual Report, 1967.
66. *IC*, 30.9.69.
67. Payne.
68. Payne.
69. 'Electricity: The Future in Scotland', 1990
70. Scottish Hydro-Electric, Annual Report, 1989-90.
71. *Northern Times*, 23.7.99.
72. *The Scotsman*, 21.7.01.

Main Sources

Association of Scientific Workers, *Highland Power* (Glasgow, 1945)

Campbell, Patrick, *Tunnel Tigers* (New Jersey, 2000)

Douglas Simpson, W., *Portrait of the Highlands* (London, 1969)

Ford, Gillean M. (compiler), *Tunnellers, Tango Dancers and Team Mates* (Stirling, 2000)

Gordon, Seton, *Highlands of Scotland* (London, 1951)

Gunn, Neil, *Off in a Boat* (London, 1938)

Johnston, Thomas, *Memories* (London, 1952)

Keillar, Ian, *Electricity in the North* (Moray Field Club, 1981)

MacColl, A. E., 'Hydro-Electric Development in Scotland, 1946', in Leith T., Broadley I., Osborn H, Robb J, eds, Institution of Engineers and Shipbuilders in Scotland, *Mirror of History: A Millennium Commemorative Volume* (Glasgow, 2000)

MacDonald, C. M., ed, *The Third Statistical Account of Scotland: The County of Argyll* (Glasgow, 1961)

MacGill, Patrick, *Children of the Dead End* (London, 1914; re-issued Edinburgh, 1999)

Mather, A. S., ed, *The Third Statistical Account of Scotland: The County of Ross and Cromarty* (Edinburgh, 1987)

Mitchell Report No. 1. The Moriston Dam Scheme (1957)

Murray, W. H., *Highland Landscape: A Survey* (National Trust for Scotland, 1962)

North of Scotland Hydro-Electric Board. First Report (Edinburgh, 1945)

——————— Loch Sloy Hydro-Electric Scheme (Edinburgh, 1950)

——————— *Power from the Glens* (Edinburgh, 1973)

Payne, Peter, *The Hydro* (Aberdeen, 1988)

Quigley, Hugh, *The Highlands of Scotland* (London, 1936)

Scottish Office. Report of the Committee on Hydro-Electric Development in Scotland. Cmd 6406 (Cooper Report) (Edinburgh, 1943)

ScottishPower, *Cruachan: The Hollow Mountain* (Glasgow, undated)

Scottish and Southern Energy plc. Annual Review and Summary Financial Statement, 2000

———— Annual Review and Summary Financial Statement, 2001.

Stevenson, James, 'The construction of Loch Sloy dam', Proceedings of the Institution of Civil Engineers, Part III (August, 1952)

Taylor, D. B., ed, *The Third Statistical Account of Scotland: The Counties of Perth and Kinross* (Edinburgh, 1979)

Thompson, Francis, *The Highlands and Islands* (London, 1974)

Turnock, David, *Patterns of Highland Development* (London, 1970)

Appendix

The major hydro-electric schemes built for the North of Scotland Hydro-Electric Board between 1945 and 1975. (The Board also operated diesel and steam power stations in Dundee, the Hebrides, Orkney and Shetland.)

SCHEME (with year of completion)	MAIN DETAILS
Loch Sloy 1950	Buttress dam on Loch Sloy (Balfour Beatty & Co.), with main tunnel and surge shaft (Edmund Nuttall Sons & Co.) leading to pipeline (Sir William Arrol & Co.) feeding a power station (Hugh Leggat Ltd) on Inveruglas Bay on Loch Lomond. Average annual output 120 million units.
1959	*Glen Shira phase*: pre-stressed gravity Allt-na-Lairigie dam (Maples, Ridgway & Partners Ltd) and tunnel to power station on River Fyne. Round-headed buttress dam on Lochan Shira Mor (A. and M. Carmichael Ltd), power station at Sronmor, concrete gravity and earth-fill with concrete core dam (A. and M. Carmichael Ltd) on Lochan Sron Mor with tunnel to Clachan power station on Loch Fyne. Average annual output: Sron Mor 6 million units; Allt na Lairigie 20 million units; Clachan 74 million units.
Tummel – Garry Pre-WWII (Grampian)	Existing structures (Grampian developments): Dam at east end of Loch Rannoch (1930). Dam at south end of Loch Ericht, tunnel to power station at the west end of Loch Rannoch (1930). Dunalastair intake dam on the Tummel river, and Tummel Bridge power station (1933). Dam at north end of Loch Ericht, tunnel from Loch Garry to Loch Ericht, dams on Loch Seilich and Loch Cuaich, tunnel between Loch Seilich and Loch Cuaich (1940). Average annual output: Rannoch 174 million units; Tummel 120 million units.
1951	*Phase 1*: Mass gravity Clunie Dam (George Wimpey & Co.) at the east end of Loch Tummel, with tunnel (Cementation Co. Ltd) to Clunie power station. Outfall was combined with the flow of the Garry to the mass gravity dam and power station at Pitlochry (Wimpey). Loch Faskally created behind the Pitlochry dam. Average annual output: Clunie 165 million units; Pitlochry 55 million units.
1955	Additional power stations: Cuaich 9 million units; Loch Ericht 11 million units.
1958	*Phase 2*: Enlargement of the Tummel catchment area by diversion of waters of the upper Garry to Loch Errochty; Trinafour diamond-headed buttress dam built at the east end of Loch Errochty (1957) (A. and M. Carmichael Ltd) and tunnel driven from Loch Errochty to Loch Tummel via the Errochty power station. Concrete gravity dam and power station at Gaur (1958) (A. A. Stuart & Sons). Average annual output: Errochty 100 million units; Gaur 19 million units.
Conon Valley 1957	*Phase 1*: Dam on Loch Fannich with tunnel to Grudie Bridge power station at west end of Loch Luichart (Balfour Beatty & Co.). Average annual output: 82 million units.

1957	*Phase 2*: Concrete gravity and earth-filled Glascarnoch Dam, and dam (earth-filled with concrete core) at Vaich. Lochs Glascarnoch and Vaich created (Reed & Mallik). Aqueducts and tunnels to feed Loch Vaich, and tunnel from Vaich to Glascarnoch. Tunnel from Glascarnoch to Mossford power station on Loch Luichart (A. and M. Carmichael). Barrage on River Bran and power station on Loch Achanalt. Concrete gravity, buttress and earth-fill dam on River Meig, which created Loch Meig, and tunnel to Loch Luichart (Duncan Logan Ltd). Mass gravity dam on Loch Luichart and tunnel to Luichart power station (Reed & Mallik). Mass gravity dam and power station at Torr Achilty, which created Loch Achonachie (William Tawse Ltd). Average annual output: Achanalt 8 million units; Mossford 112 million units; Luichart 124 million units; Torr Achilty 36 million units.
1961	*Phase 3*: Twin mass gravity dams on River Orrin created reservoir, and tunnel to Orrin power station on Loch Achonachie (Duncan Logan Ltd). Average annual output: 76 million units.
Affric – Cannich 1952	Mass gravity dam on Loch Mullardoch, and tunnel to Loch Benevean. Tunnel from Loch Benevean to Fasnakyle power station (John Cochrane & Sons Ltd). Average annual output: Mullardoch 8 million units; Fasnakyle 223 million units.
Strathfarrar – Kilmorack 1963	Arch constant angle dam on Loch Monar, and tunnel to Deanie power station (Mitchell Construction Co.). Mass gravity dams at Loichel (on Loch Monar) and on Loch Beannachran, and tunnel to Culligran power station (Duncan Logan Ltd). Mass gravity dams with power stations at Aigas and Kilmorack (A. A. Stuart & Sons Ltd). Average annual output: Deanie 94 million units; Culligran 57 million units; Aigas 55 million units; Kilmorack 55 million units.
Moriston – Garry 1957	*Garry*: Rockfill with articulated concrete face dam on Loch Quoich to create reservoir, and tunnel to Quoich power station on River Garry (Richard Costain Ltd). Mass gravity dam on Loch Garry and tunnel to Invergarry power station (Whatlings Ltd). Average annual output: Quoich 77 million units; Invergarry 72 million units. *Moriston*: Mass gravity dams on Lochs Loyne and Cluanie, with tunnel from Loch Loyne to Loch Cluanie, and tunnel from Loch Cluanie to Ceannacroc power station (Mitchell Construction Co.). Mass gravity dam at Dundreggan and Glenmoriston power station, with tailrace tunnel to Loch Ness (Duncan Logan Ltd). Aqueduct system to feed Livishie power station (Duncan Logan Ltd). Average annual ouput: Ceannacroc 73 million units; Livishie 27 million units; Glenmoriston 114 million units.
Loch Shin 1960	Concrete gravity and earth-fill dam and power station at Lairg. Tunnel from Little Loch Shin to Inveran power station. Duchally weir, aqueducts and tunnels to divert headwaters of Cassley River to Cassley power station on Loch Shin (George Wimpey & Co. Ltd). Average annual output: Cassley 24 million units; Lairg 10 million units; Shin (Inveran) 103 million units.
Breadalbane 1961	*Lawers section*: Massive buttress Lawers dam on Lochan na Lairige, aqueducts and tunnels to enlarge the lochan's catchment area, tunnel to Finlarig power station on Loch Tay (Cementation Co. Ltd). Average annual output: 64 million units. *St Fillans section*: rockfill dam with upstream concrete facing on Loch Breaclaich (R. J. McLeod Ltd) with tunnel to Lednock power station on Loch Lednock. Diamond-headed butress dam on Loch Lednock (Taylor Woodrow Ltd). Tunnel to St Fillans power station on Loch Earn. Tunnel from Loch Earn to Dalchonzie power station. (Tunnels: Mitchell Construction Co.) Average annual output: Lednock 5 million units; St Fillans 76 million units; Dalchonzie 18 million units.

	Killin section: Tunnel, and aqueducts, and massive buttress dam and power station at Lubreoch on Loch Lyon (James Miller & Partners Ltd). Average annual output: 13 million units. Dam on Loch Giorra (Edmund Nuttall Sons & Co), tunnel to Cashlie power station. Dam on Stronuich reservoir and tunnel to Lochay power station (Edmund Nuttall Sons & Co). Average annual output 25 million units; Lochay 160 million units.
Awe/ Ben Cruachan 1965	Pumped storage scheme: Dam on Ben Cruachan reservoir and underground power station in Ben Cruachan (Edmund Nuttall Sons & Co.). Installed capacity 400 megawatts. Associated conventional schemes: Dam on Loch Nant, with aqueducts and tunnel to Nant power station on Loch Awe. Dam on north-west arm of Loch Awe, with tunnel to Inverawe power station on Loch Etive. Average annual output: Nant 27 million units; Inverawe 100 million units.
Foyers 1975	Two dams on Loch Mhor, aqueduct to divert River Fechlin to Loch Mhor, with tunnel to power stations of Foyers Falls and Foyers on Loch Ness (Duncan Logan Ltd – completed by Edmund Nuttall Ltd). Pumped storage capacity 300 megawatts; average annual output: Foyers Falls 6 million units.
SMALL SCHEMES	
Morar 1948	Dam and power station on the River Morar. Average annual output 3 million units.
Lochalsh 1948	Dam on the Allt Gleann Udalain and power station at Nostie Bridge. Average annual output 6 million units.
Mucomir Cut 1951	Power station installed in a channel between the southern end of Loch Lochy, close to the Caledonian Canal, and the Spean River. Average annual output 9 million units.
Striven 1951	Dam on Loch Tarsan and power station at the head of Loch Striven, near Kyles of Bute. Average annual output 22 million units.
Kerry Falls 1952	Reservoir at Loch Bad an Sgalaig and pipeline on the River Kerry, Gairloch, with power station. Average annual output 5 million units.
Storr Lochs 1952	Dam and power station on the north-east coast of Skye. Average annual output 7 million units.
Glen Lussa 1952	Power station on Glen Lussa Water draining Loch Lussa, Kintyre peninsula. Average annual output 10 million units.
Loch Chliostair 1954	Harris: dam, pipeline and power station on Eaval River. Average annual output 3 million units.
Loch Dubh 1955	Dam, pipeline and power station in Strath Kanaird, Ullapool. Average annual output 5 million units.
Kilmelfort 1956	Dams on Lochs Tralaig and na Sreinge, and on the Melfort Pass near the Ardrishaig-Oban road, feeding power station on the River Oude. Average annual output 11 million units.
Loch Gair 1960	Dam on Loch Glashan, north-east of Lochgilphead, with pipeline and tunnel to power station on Loch Gair. Average annual output 18 million units.
Loch Gisla 1960	Gisla River drainage system, with power station at Gisla, draining into Little Loch Roag, south-west Lewis. Average annual output 2 million units